SPACE COLOR ART

空间彩色艺术

Modern Office Color Schemes
现代办公空间色彩搭配

君誉文化 策划
高迪国际出版有限公司 编
赵翔宇 邹红 周双 赵远 译

大连理工大学出版社

Preface 1

序言1

Offices now are not merely a place to work with all necessary work equipment, but a space to do something productive, innovative, a place to think and for such work related brainstorming; activities indirectly related to working such as leisure, exercising and enjoying also have become an integral part of the work process.

Therefore, task for architects today is creating offices which are not just functional and comfortable but also stimulate work and creativity.

Whether it is music you need or total silence, connected to outdoor or enclosed, colorful or subtle in tones, you have to be able to create that ideal environment to be at your most productive.

Role of architect and designers for the creation of spaces for work is especially important, because a designer is the only person who can add aesthetic and creative constituents to a new office. The client is the person who may only tell you the requirements for his office space. You as a designer have to connect to his way of working and understand his psychology, and at the same time the taste of the client which will help you to generate response towards the design challenge.

Most important features for designing an office space, to name a few are; proper lighting – comfortable and appropriate lighting is a must for the workers to concentrate and work with feeling stressed and good or even some amount of natural lighting would be a bonus. Connectivity – Over the years modern offices have changed dramatically and now the concept of closed cubicles have converted to multiple connected spaces, open areas, breakout zones where the users can interact, enjoy, exchange ideas and be more productive. Colors – Colors to a great extent impacts on the moods and feelings and this directly affects the quality of work and creativity. Thus the colors and tones of the spaces and the elements should be decided in a way that it enhances the working environment.

Overall, the challenge is not only to create beautiful, vibrant colorful spaces...It is not to achieve balance between open and closed, between private and collaborative, but to create a space that is so great that people would like to come simply because there is no better place to do what they need to do. People come to a great workplace because they want to and not because they have to.

The book further is a collection of such beautiful offices which will reflect the innovations and responses by the creative people around the world. We hope this will inspire you to create more surprising and creative projects.

Kapil Aggarwal
The Principle of Spaces Architects@ka
卡皮尔·阿加沃尔
Spaces Architects@ka 设计事务所董事

办公室现在不仅仅是一个用必要的工作设备进行工作的地方，而是用来做一些富有成效和创新事情的空间，也是用于思考和进行和工作相关的集思广益的地方。和工作息息相关的一些活动，例如，休闲、锻炼和娱乐也已经成为工作过程中不可分割的一部分。

因此，当今建筑师的任务不仅仅要建造功能型和舒适型的办公室环境，也要让办公环境促进工作和激发员工的创造力。

无论是需要来点音乐还是希望保持安静、希望连接到室外还是完全封闭、色调华丽还是细致，你必须能够尽你所能营造出理想的办公环境。

建筑师和设计师在创建工作空间上起着重要的作用，因为只有设计师能够将美学元素和创造性添加到新的办公室中。客户可能只会告诉你他对办公室空间的要求，作为一个设计师，你必须了解客户的工作方式、客户的心理，以及品位，从而帮助你克服设计中的难点。

办公室空间设计最重要的特点有：适当的照明——舒服而适宜的灯光对于员工集中精力和紧张地工作是必需的。良好的自然光，哪怕不多，也会锦上添花；连通性——多年来现代办公室设计已经有了巨大的变化，现在封闭隔间的概念已经转换为多个相连空间、开放领域和突破区域，用户可以互动、娱乐、交流思想并使工作更有效率；颜色——颜色在很大程度上影响心情和感受，这直接影响到工作的质量和员工的创造力，因此空间和元素的颜色和色调的选择在某种程度上可提升工作环境。

总体来说，设计面临的挑战不仅是创造美丽、活泼、多彩的空间；不是在开放和封闭、私密和公共空间之间达到平衡，而是要营造一个无与伦比的空间，人们愿意来到这里仅仅是因为他们在这里可以做想做的事情。人们来到这个理想的工作场所是因为他们愿意来而不是因为他们必须来。

而且这本书收集了很多漂亮的办公室设计，反映了世界各国创新型人才的创造力和反应力。我们希望这会激发你去创造更多优秀而新颖的项目。

Preface 2

It has become an inescapable reality: The profound digital, structural change of our world calls for a reinvention of how we organize and design our working environments.

The miniaturisation of our digital working tools leads to a multitude of consequences: Work stations become smaller, our work more mobile. Today many of us are no longer bound to our desks. Instead we can work in different places – in the department, the building, in a coffee shop, at the airport. Moreover the ubiquity of telecommuting and the home office have made the workday much more flexible. Some companies have responded to these changing conditions by doing away with the traditional allocation of individual work stations in favour of a non-territorial desk system. And because not every employee is physically present in the office at all times, a reduced number of work stations will suffice. Actual space requirements are reduced further still as technology continues to shrink in size and employees limit the amount of physical things they require to ensure greater mobility. Spaces are thus employed in a much more efficient way. At the same time, designing the work station to cater to each respective user in an individually adaptable and scalable way is an important challenge here. All this changes how we conceive the built environment of work. On one hand space can be used more efficiently, which results in many fantastic responses from the real estate side. On the other hand we need to develop new typologies that suit the changing demands of our work.

Our new work organisation means the transparent linking of knowledge with people, computers, machines and the respective environment – across departments, companies or even work places. Cross-departmental projects require much greater exchange and organisation. Creative interdisciplinary setups, with team members coming from different disciplines, departments, locations, agencies or working with external specialists are designed to concentrate specialist knowledge from different contexts. This ceases to be easily feasible at one location and on predetermined dates.

In this respect, interior layout and design plays a pivotal role, which can initiate significant developments. So instead of thinking about organizing desks, the task now shifts to creating opportunities for chance encounters. Generous tea kitchens, the design of seating niches around circulation areas and the expansion of large offices with multi-functional zones are spatial answers that are currently being employed. The activity of a knowledge worker varies between focused work, empathic customer meetings, creative solution finding, provisional prototyping, regenerative relaxation, as well as reaping the possibilities of inspiration. To meet these different demands on his way of working and thinking, the knowledge worker will look for rooms that best put him in the desired work attitude. So in future we need less standardised and more varied environments for flexible ways of working – rooms that associatively and emotionally transfer us into the respective working mode.

Above and beyond that, working environments are a part of Corporate Identity. They represent a company's culture and identity to the outside and inside world. This is even more the case as most companies are fighting hard to acquire the best talents on the market. So the environment becomes a critical asset in the acquisition of

Gunter Fleitz and Peter Ippolito
Managing Partners of Ippolito Fleitz Group – Identity Architects
甘特·福莱兹、彼得·伊波利托
伊波利托·福莱兹集团任事股东

human resources.

With such changes to the working environment, it becomes more important than ever to respect the sensitivities of employees. Because ultimately, employee satisfaction is one of the key assets of any company. Offering communication and recreation landscapes is one thing, but factors such as light, indoor climate, acoustics and discretion are also important comfort criteria.

The real challenge in interior design is to respond to changing working conditions as well as to the increased expectations of employees, and to transform modern working environments into places that shape identity.

This book takes you on a voyage of discovery to some of the most stunning office concepts from all around the world.

这个世界深奥的数字化结构变化要求我们重新组织和设计我们的工作环境，这已经成为一个无法逃避的事实。

数字工作工具的微型化导致一系列后果：工作站变得更小，工作更不固定。现在我们很多人不再被束缚于办公桌周围，而是可以在不同的地点工作——在部门里、楼里、咖啡屋、机场。另外，远程办公的普遍盛行和在家办公使工作时间更加灵活。一些公司通过将传统的个人工作站转变为无界限划分的办公系统来应对这些环境的改变。因为不是每个员工都时时刻刻呆在办公室，所以虽然工作站的数量已经减少，但是也足够了。科技规模继续萎缩，员工所需要的物件的减少确保了更大的流动性，实际的空间需求进一步减少，空间被更有效地利用。同时，为迎合每一个用户的需求，用一种独特的、适应性强并可扩展的方式设计工作站是一个很大的挑战，而这一切改变着我们对建造工作环境的想法。一方面，空间可以被更有效地利用，地产商对此也非常认同，另一方面，我们需要发展新的科技，从而来适应日益变化的工作需求。

我们新的工作组织意味着将知识和人、计算机、机器以及各自的环境之间跨部门、公司甚至工作场所进行透明链接。跨部门项目要求更多的交流和组织。新颖且跨领域的机构设置和来自不同学科、部门、地点、机构以及和外部专家共同工作的队员可以集中不同情景下的专业知识，而这在某一地点或特定的日期内实现是不容易的。

在这方面，室内布局和设计起着重要作用，它可以推动重大的进展。所以现在的任务转向创造偶遇的机会，而不是考虑设置办公桌。宽敞的茶餐厅、流通领域桌椅的设计以及多功能办公区域的扩展是现在正在使用的方案。脑力工作者的活动在专心工作、投入的客户会议、创新的解决方案、暂定的原型设计、更新的消遣方式以及获取灵感方面存在着差异。为了满足工作和思考方式上的不同需求，脑力工作者会寻找一个能使他们处于最好工作状态的办公空间，所以未来我们需要不太统一的，多样的工作环境来适应灵活的工作方式，这样的办公房间组合起来从情感上使我们转移到各自的工作模式中。

除此之外，工作环境是企业形象的一部分。它对内对外都代表着一个公司的文化和身份。在大多数公司激烈地争夺市场人才的情况下，更是如此。所以工作环境在获取人力资源方面是一个关键因素。

随着工作环境的变化，现在比以往更尊重员工的敏感性。因为，最终员工满意度对任何公司来说都是宝贵财产。提供交流和休闲的景观环境是一方面，但其他因素，例如，光线、室内温度、隔声性和空间灵活度，也是评判舒适度的重要指标。

室内设计真正的挑战是应对不断变化的工作环境和员工期望值的增加，以及将现代工作环境转换为塑造身份的地方。

这本书将带你一道旅行，去发现一些来自世界各地的绝妙的办公室设计理念。

Contents

ASTRAL MEDIA 星际传媒	010	**GOOGLE MADRID HQ** 谷歌马德里总部	086
OFFICE SITEGROUND SiteGround 办公室	020	**MASISA OFFICE** Masisa 办公室	100
A BUSINESS IN HERON TOWER 苍鹭大厦商业空间	028	**OCA GROUP** OCA 集团	110
INFOR GLOBAL SOLUTIONS Infor Global Solutions 办公空间	036	**URALCHEM HEADQUARTERS** Uralchem 总部	120
YARRA VALLEY WATER 亚拉谷溪水	044	**RED BULL NORTH AMERICA** 北美红牛	128
AKER SOLUTIONS 阿克工程集团设计	056	**CTAC'S – HERTOGENBOSCH** CTAC 斯海尔托亨博斯办公室	136
ALCATEL LUCENT HEADQUARTER 阿尔卡特朗讯总部设计	064	**OFFICE SPACE IN TOWN – @WATERLOO** Office Space in Town – @ 滑铁卢	144
FLEETPARTNERS FleetPartners 办公区	078		

SITEGROUND SiteGround	154
URBIS SYDNEY 悉尼乌尔比斯	160
DREES & SOMMER STUTTGART Drees & Sommer 斯图加特办公室	168
HSB HSB	178
INNOCEAN HEADQUARTERS EUROPE 伊诺盛欧洲总部	194
YANDEX SAINT PETERSBURG OFFICE – 4 圣彼得堡 Yandex 办公室 - 4	204
ALFA BANK OFFICE 阿尔法银行办公室	214
DROGA5 Droga5 办公空间	228
GREENLAND CENTER 绿地中心	236
LEASEPLAN LeasePlan 办公空间	244
MICROSOFT SPAIN HEADQUARTERS 微软西班牙总部	252
NOTTING HILL 诺丁山	262
ONEFOOTBALL Onefootball 办公室	270
COMPULSIVE PRODUCTIONS 令人着迷的办公空间	280
CUBIX OFFICE Cubix 办公室	288
SPACES ARCHITECTS@KA OFFICE Spaces Architects@ka 办公室	296
THE BRIDGE 桥	308
INDEX 索引	316

ASTRAL MEDIA RELOCATED APPROXIMATELY 350 EMPLOYEES TO FOUR FLOORS IN THE HEART OF THE ACTION IN DOWNTOWN MONTREAL.

ASTRAL MEDIA

星际传媒

*Designer*_Annie Desrochers, Chantal Ladrie (Project Manager), Marie-Elaine Globensky (Graphic), Véronique Richard (Industrial and Graphic), Sandra Neill (Associate) / *Design Company*_Lemay / *Project Team*_Isabelle Matte, François Descôteaux, Caroline Lemay *Partner in Charge*_Louis T. Lemay / *Technician*_Leonor Oshiro, Phary Louis-Jean Technicians, Serge Tremblay, Geneviève Telmosse *General Contractor*_Patella inc. / *Engineer*_Planifitech Inc. (Electro-Mechanical), Nicolet Chartrand Knoll Limitée (Structural/Civil) *Construction Cost*_$10M / *Client*_Astral Media / *Location*_Montreal, Canada / *Area*_6,000 m² / *Photographer*_Claude-Simon Langlois

The goals of this large-scale project included elaborating new furniture standards, fitting up flexible meeting spaces and optimizing employee interconnectivity in a contemporary, energetic and versatile working environment.

Based on the client's four different business units (radio, television, advertising and digital media), our concept was inspired by key broadcasting industry words such as influence, communication, movement and exchange. The concept plays on the contrast between the medium and the message and manifests itself by means of undulating and pixelated graphic interventions.

The project's main challenge consisted in guiding the client's transition towards a youthful and modern image and accommodating a large number previously dispersed employees in a single open and standardized space.

Aside from workstations, we fitted up a main reception area, various meeting spaces (conference and meeting rooms, agora, etc.) as well as common services (dining room, lounge, café, copy centre) on each floor. In order to create a rhythm and a gradation throughout the playful 6,000 m² space, each floor was identified with its own color and the levels were linked by a central glass staircase.

Astral Media's new flexible, functional and bright premises provide a human working experience and favour communication between employees.

COLORS MATCH
色彩搭配

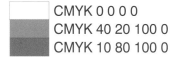

CMYK 0 0 0 0
CMYK 40 20 100 0
CMYK 10 80 100 0

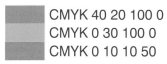

CMYK 40 20 100 0
CMYK 0 30 100 0
CMYK 0 10 10 50

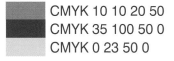

CMYK 10 10 20 50
CMYK 35 100 50 0
CMYK 0 23 50 0

CMYK 50 50 50 0
CMYK 40 20 100 0
CMYK 10 80 100 0

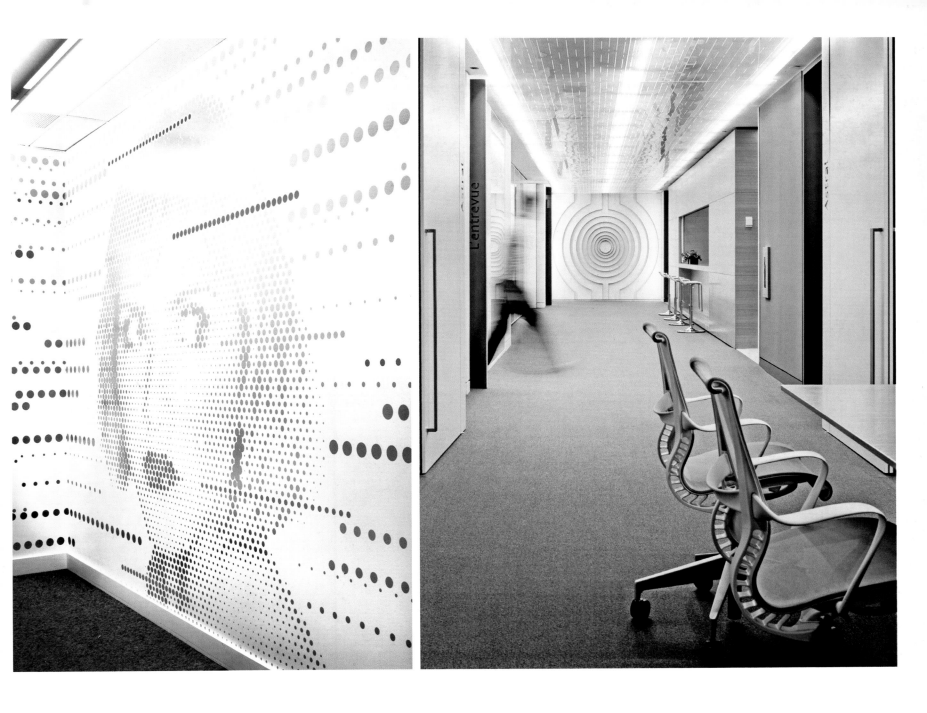

星际传媒把约350名员工迁移到位于蒙特利尔市中心区域的四层办公楼中。这个大型项目的目标包括详细制定新的家具标准，装修出灵活的会议空间以及优化员工在现代化、充满活力和多样性的工作环境中的互动性。

基于客户的四个不同的业务部门（广播、电视、广告和数字媒体），我们的概念灵感来自于广电行业的关键词语，如影响力、沟通、运动和交流。这一理念运用于媒体和信息之间的对比，并且通过起伏和滤镜效果的图形展现概念本身。

该项目的主要挑战在于指导客户朝着年轻和现代的形象转变，并在一个开放的、标准化的空间中容纳大量之前座位分散的员工。

除了工作站，我们还配备了一个主要的接待区、各种会议室（重大会议室和小会议室，集会厅等），以及各楼层均有的公共服务区（餐厅、休息室、咖啡厅和复印中心）。为了使这6000平方米的场地更有层次感和节奏感，每个楼层都涂有不同的颜色，并且通过一个中央玻璃楼梯相连接。

星际传媒这个新颖灵活、功能齐全，并且明亮的场所，为员工提供了人性化的工作体验，并有助于同事之间的交流。

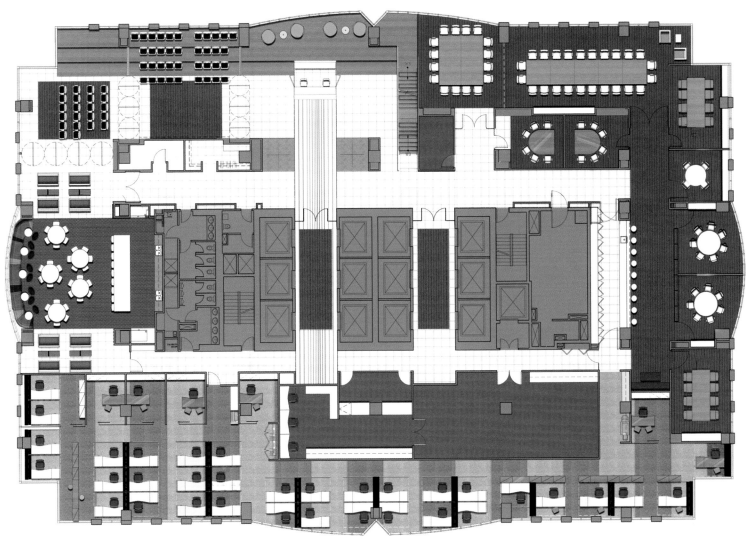

OFFICE SITEGROUND

Designer_Funkt Architects / Location_Bulgaria / Area_1,600 m²
Photographer_Asen Emilov, Vencislava Vassileva

SiteGround 办公室

SITEGROUND IS A FAST DEVELOPING COMPANY IN INFORMATION TECHNOLOGY SPHERE WITH CONSTANTLY GROWING NEEDS FOR SPACE.

Their wish to create a fun, personalized and motivating work environment leads them to our team. The assignment is to design a great space where entertainment and relaxation are in balance with the work flow. The office is located in a new office building in the business area of Sofia and occupies an entire floor. After exploring the way company's employees use the office, we come to the idea of a fluid open space with special collective hubs — for both work and relaxation. The newly created space combines various zones with different character — parlor, lounge, conference rooms, relaxation rooms, gaming, working, kitchen, cloakroom, manager offices, etc. For four months we converted an empty floor into cheerful and poetic space where working hours from 9 to 5 is no longer a principle, work and fun are complementary and people feel special.

Colorful paper models create visual illusions of embracing forms on the walls and floor. The folded volumes in the reception area continue such a story line. Rhombs and triangles reappear in shelves and flooring patterns in order to integrate the whole space. The lounge area is multifunctional — for meetings, resting, reading or work, located close to the reception and is visually connected with the meeting rooms. All conference walls are converted into whiteboards. Besides standard office equipment we introduce a table tennis, pool, football, and number of other games. Custom — made wallpaper and carpet patterns in pop art style together with fluidly dispersed furniture create color codes for the different themes.

What is very important for the client's approach toward software, is the custom-made production from "a" to "z" for every new project. This idea translates into a handmade craft approach for all the different elements custom-made for this company. Handmade paper shapes is not only another way to have fun and relax but also something that residents continue to make and relate to their work on daily bases.

COLORS MATCH
色彩搭配

CMYK 60 0 100 0
CMYK 0 0 0 0
CMYK 0 0 0 45
CMYK 15 8 5 60

CMYK 0 23 80 0
CMYK 17 0 80 0
CMYK 0 0 0 30
CMYK 0 0 0 15

CMYK 18 82 0 0
CMYK 31 44 5 2
CMYK 0 0 0 90

CMYK 95 25 15 0
CMYK 0 0 0 0
CMYK 15 10 0 40

CMYK 0 0 0 0
CMYK 8 8 15 0
CMYK 15 38 60 0
CMYK 56 15 80 0

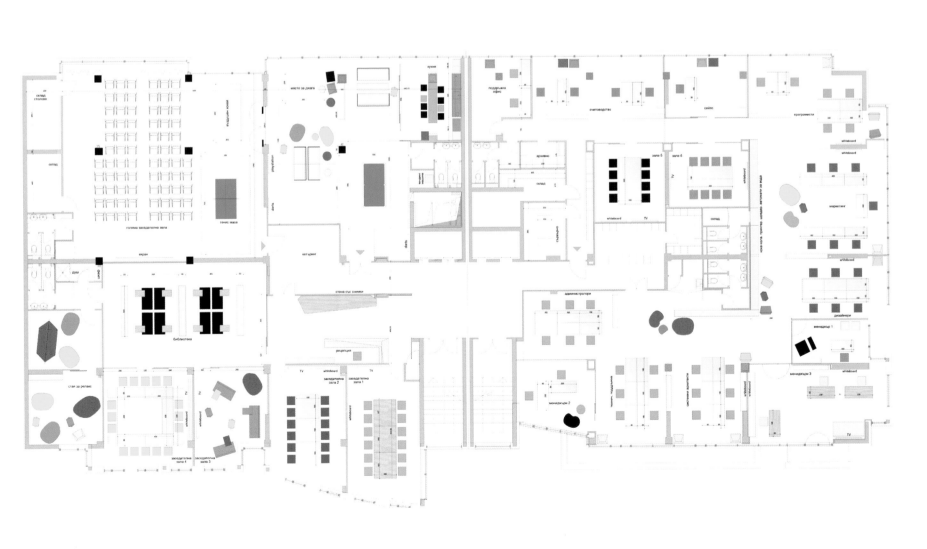

SiteGround 是在信息技术领域飞速发展的公司，因此对空间的需求日益扩大。该公司想营造一个集趣味性、个性化和激发员工动力为一体的工作环境，为此找到了我们团队。这项任务是要设计一个能平衡娱乐、休闲与业务流程的空间。该办公室位于索菲亚商业区的一栋新办公楼里，占该楼的一整层。在了解了公司员工的办公方式后，我们决定设计一个流动的开放式空间，该空间配有独特的兼工作休闲为一体的聚集中心。新建空间的不同区域各有特点，包括会客厅、休息室、会议室、放松室、游戏区、工作区、厨房、衣帽间以及经理办公室等。四个月的时间里，我们把一个空荡荡的楼层改变成了令人愉快又富有诗意的空间，在这里朝九晚五的工作时间不再是原则性问题，工作和娱乐互为补充，让人感觉很特别。

墙和地板上色彩丰富的纸模型营造了一种环绕式的视觉幻想。接待区里折起的画卷继续着这样的故事情节。为了使整体空间融合在一起，架子和地板的图案均呈菱形和三角形。休息区在接待室旁边，在视觉上与会议室相连；它功能繁多，可以开会，休息，阅读，也可以工作。所有会议室墙壁都是白色书写板。除了标准的办公室设备，我们还引入了乒乓球、撞球、足球和其他一些游戏。定制的墙纸和波普艺术风格的地毯图案与流动、分散放置的家具营造出不同主题的颜色代码。

为每一个新项目都从头到尾量身定做是非常重要的。正是由于这样的工作理念，对此公司定制使用的各种元素我们都采用了手工工艺方法。手工纸模型不仅是一种娱乐和放松的方式还是员工每天继续要做的、并且与工作相关的事。

A BUSINESS IN HERON TOWER

苍鹭大厦商业空间

Designer_Peldon Rose / Duration_13 Weeks / Location_London, UK
Area_1,077.67 m² / Photographer_Matthew Beedle

AS A UNIQUE BUSINESS THAT ENCOURAGES ITS PARTNERS TO DEVELOP THEIR OWN ENTERPRISES AND BRANDS, OUR CLIENT ENTRUSTED THE CREATION OF AN OUT-OF-THE-ORDINARY SPACE TO US.

After securing a premium floor in London's iconic Heron Tower, we started creating the interior. Our client was open to ideas, as long as the reception area made a dramatic first impression and doubled up as a hub.

So we added to the angled walls with contrasting materials like white marbled floors, ceilings clad with timber planks, and tessellated textures, patterns and tiles for the walls. A mirrored wall bounces glittering reflections off the reception into the wider space and forms a separation of the open coffee bar beyond.

By centralizing the reception area and adjoining coffee bar, we were able to make the most of this sought-after space, with corridors running off the central hub and leading to striking meeting rooms positioned around the outside. With the powerful geometric theme leading throughout.

COLORS MATCH
色彩搭配

CMYK 0 0 0 0
CMYK 15 38 70 0

CMYK 8 10 15 50
CMYK 0 0 0 0
CMYK 15 38 70 0

CMYK 100 60 20 0
CMYK 5 5 5 2
CMYK 10 35 65 0

作为一家鼓励合作伙伴开创其自己的事业和品牌的独特企业，我们的客户委托我们创造一个卓越非凡的空间。

在伦敦地标性的苍鹭大厦选定一个绝佳的楼层后，我们便开始其室内设计。只要接待处能够给人留下引人注目的第一印象，并且能够同时成为活动的中心位置，客户愿意接受任何开放的想法。

因此，我们在有棱角的墙面中加入了对比强烈的材料，如白色的大理石地板，天花板镶上了木质板材，并用小块大理石镶嵌出纹理、图案，并在墙上贴上了瓷砖。镜面墙壁可以反射接待室并且显示出更大的空间，同时也间隔了另一边的开放式咖啡吧。

由于接待区位于中央位置并毗邻咖啡吧，我们能够充分利用这一抢眼的空间，廊道围绕着中心轴，通往引人注目的外部会议室，强大的几何主题贯穿着始终。

INFOR GLOBAL SOLUTIONS

Infor Global Solutions 办公空间

Designer_VOA Associates Incorporated / Project Leader_Leonard M. Cerame
Location_New York, NY, U.S.A / Photographer_Ari Burling

"THE DESIGN OF INFOR'S NEW HEADQUARTERS IS AN INTEGRAL PART OF THEIR CORPORATE STRATEGY, DISPLAYING THE INFOR BRAND AND CAPTURING THE COMPANY'S FORWARD-THINKING WORKPLACE STYLE.

COLORS MATCH
色彩搭配

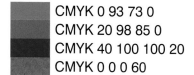

CMYK 0 93 73 0
CMYK 20 98 85 0
CMYK 40 100 100 20
CMYK 0 0 0 60

CMYK 21 21 25 1
CMYK 38 70 80 30
CMYK 30 30 30 50

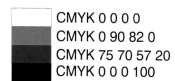

CMYK 0 0 0 0
CMYK 0 90 82 0
CMYK 75 70 57 20
CMYK 0 0 0 100

The goal was to create a workplace that fosters impromptu encounters through an open office setting while also providing ample collaborative spaces". - Leonard M. Cerame, Managing Principal

Infor, the third largest business enterprise software company in the world, relocated its headquarters and customer sales center to New York. The space was designed to capture Infor's forward-thinking workplace style, creating an office that fosters impromptu encounters through an open plan setting with ample teaming and collaborative areas. Upon entering the 4th floor reception, visitors are greeted by a two-story media wall – the first of many media screens throughout the office – creating an immersive digital experience. Two double – height floor openings create visual destination points and a spatial connection between the conference center below and the open office above.

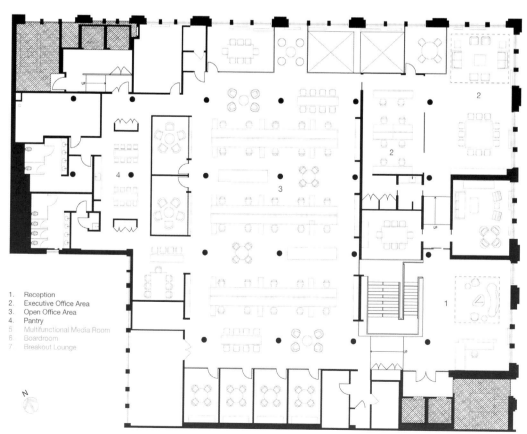

1. Reception
2. Executive Office Area
3. Open Office Area
4. Pantry
5. Multifunctional Media Room
6. Boardroom
7. Breakout Lounge

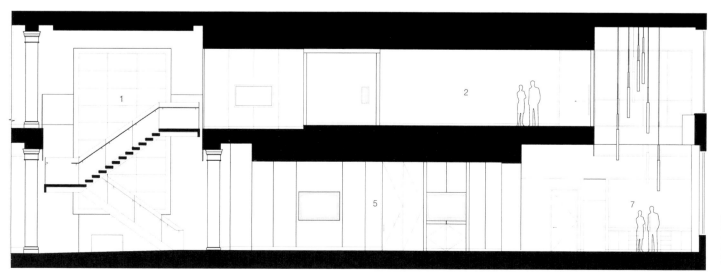

1. Reception
2. Executive Office Area
3. Open Office Area
4. Pantry
5. Multifunctional Media Room
6. Boardroom
7. Breakout Lounge

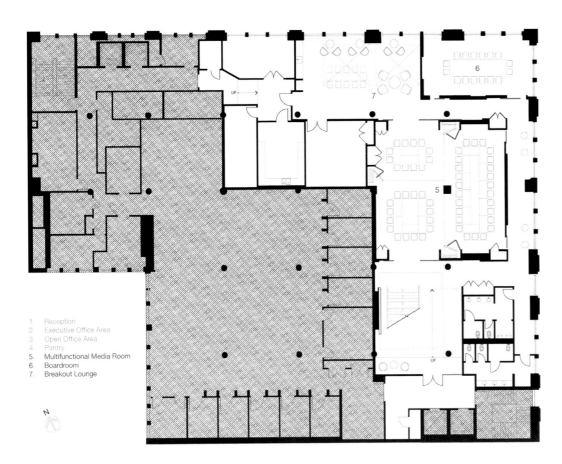

1. Reception
2. Executive Office Area
3. Open Office Area
4. Pantry
5. Multifunctional Media Room
6. Boardroom
7. Breakout Lounge

　　Infor 的新总部大楼设计是展示 Infor 品牌和体现其公司先进工作环境风格的重要组成部分。管理负责人 Leonard M. Ceremas 说道："这样设计的目的是让人们在一个宽敞而利于合作办公的环境里工作并激发人们的灵感。"

　　Infor 是世界第三大商务软件公司，它将总部和销售中心迁移至纽约。这一新总部大楼空间的设计意在抓住 Infor 前卫的工作环境风格，通过设置宽敞的合作和协作区域，使员工在不经意间邂逅。进入到第四层的接待区后，来访客人迎面看到的是两层高的多媒体墙——第一面覆盖办公室的多媒体墙——让人们沉浸到数字体验中。两个两层高的入口吸引了人们的视线，处于下方的会议中心和上方的开放办公区隔空相连。

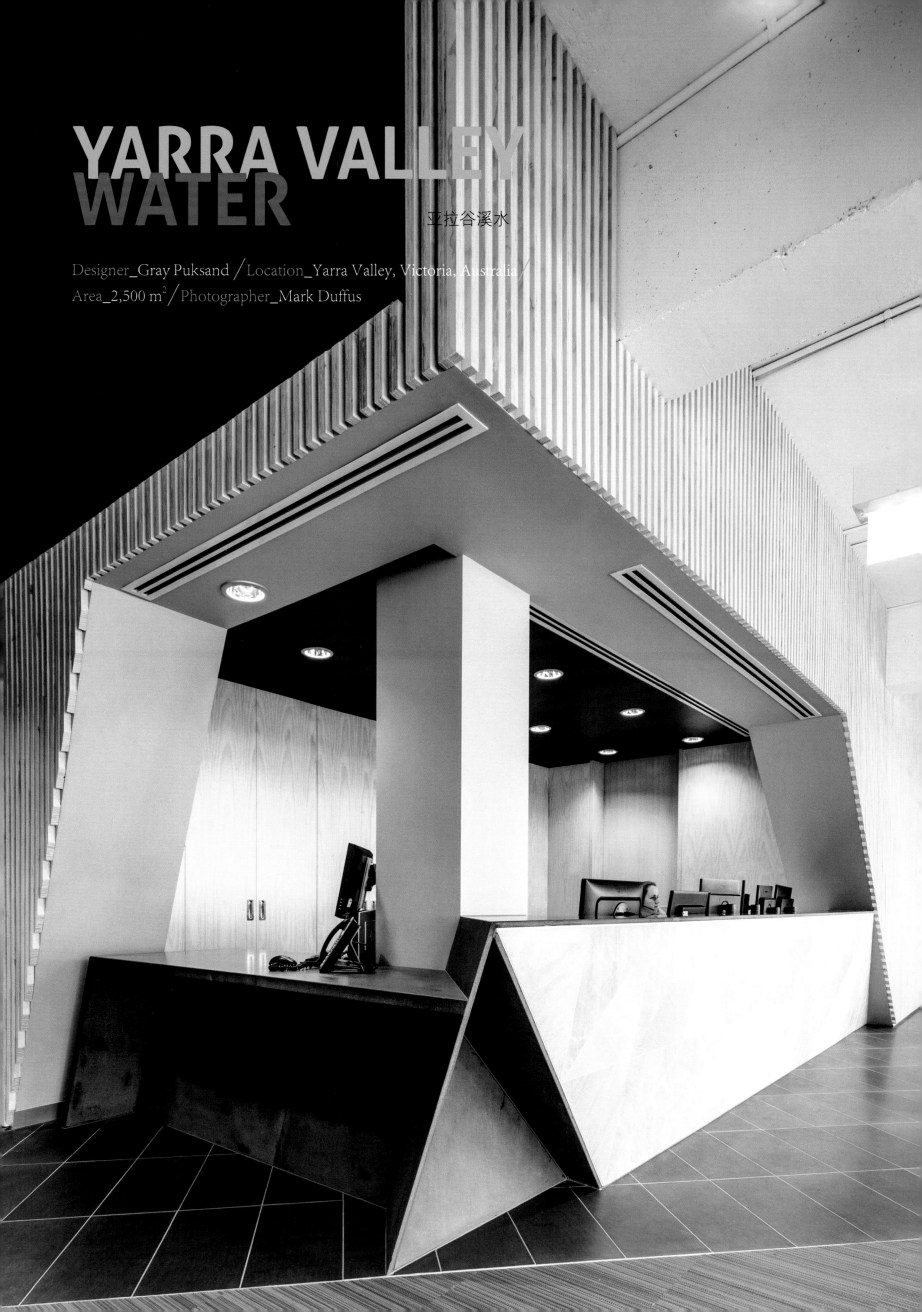

YARRA VALLEY WATER

亚拉谷溪水

Designer_Gray Puksand / Location_Yarra Valley, Victoria, Australia / Area_2,500 m² / Photographer_Mark Duffus

THE ENTIRE CAMPUS UPGRADE WAS COMPLETED OVER FOUR STAGES, EXTENDING THE BUILDING LIFE BY 40 YEARS.

Extensive work shopping derived the following aspirations; provide a collaborative working and learning environment, provoke a sense of pride and ownership, a connection with nature and an environment that evolves with the workforce.

To meet these aspirations, four zones were established and linked through planning via sight lines and circulation — Think, Chill, Meet and Focus.

The essence of the project responded to Yarra Valley Water only. It celebrates what the client does. We were inspired by the client operations; they harvest, filter and supply water. We studied how they achieve this. They harness nature, including reeds, pebbles and grasses. The detailing throughout the space is inspired by these items.

Reeds (represented by plywood slats and sun shading), pebbles through the floor finishes, and grasses in the breakout spaces. This theme continues through to graphic design and furniture selection, which was inspired by the evident color of the light spectrum refracting through water.

COLORS MATCH
色彩搭配

CMYK 0 0 0 0
CMYK 38 12 70 0
CMYK 54 30 63 0

CMYK 0 0 0 0
CMYK 0 15 25 0
CMYK 15 100 0 0
CMYK 100 95 20 0

CMYK 0 0 0 0
CMYK 93 20 12 0
CMYK 15 28 55 0

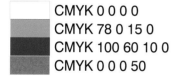
CMYK 0 0 0 0
CMYK 78 0 15 0
CMYK 100 60 10 0
CMYK 0 0 0 50

CMYK 12 22 60 0
CMYK 8 7 5 0
CMYK 55 45 40 0

整个园区升级完成要经过四个阶段,将延长建筑寿命 40 年。

大量的产品购置希望能实现以下几点期望:提供一个协作的工作和学习环境、激发自豪感和主人翁意识、紧密贴近自然以及使人力资源和环境共发展。

为实现这些期望,建立了四个区域并通过视线和循环的方式将其连在一起,这四个区域分别是:思考区、冷却区、遇见区,以及专注区。

项目的本质和亚拉谷溪水相呼应。秉承"客户为上帝"的原则。我们受客户操作的启发,他们收集、过滤和供水,而我们研究了他们的工作过程。他们利用大自然要素,包括芦苇、鹅卵石和草地,因此贯穿空间的细节设计都受到了这些东西的启发。

芦苇(以胶合板板条和太阳阴影为代表),通过地板抛光剂抛光的鹅卵石,休息区的草坪,这些主题一直贯穿于绘图设计和家具的甄选中,而这种灵感来源于经水面反射后折射出的光色。

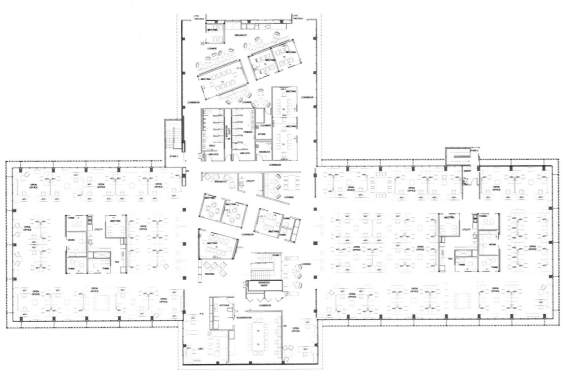
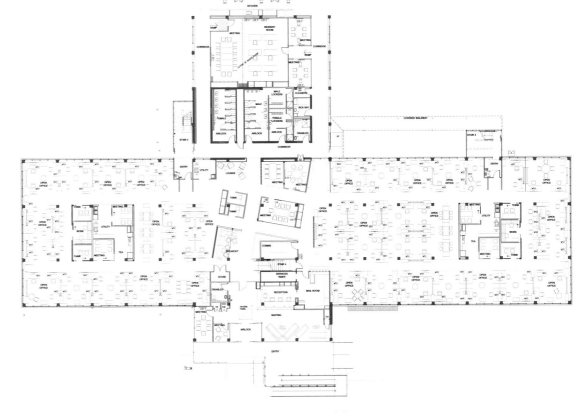

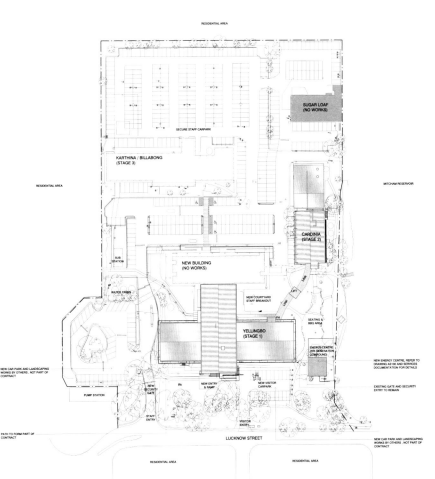

AKER
SOLUTIONS

阿克工程集团设计

Designer_Peldon Rose / ***Builder***_Peldon Rose
Build Duration_1 year (5 phases) / ***Location***_London, UK
Area_19,695 m² / ***Photographer***_Matthew Beedle

PELDON ROSE REALLY STRUCK OIL WITH THEIR LATEST DESIGN AND BUILD PROJECT FOR AKER SOLUTIONS.

Aker Solutions — an oil and gas services company — were looking to expand into a space that would attract a new generation of engineers and allow them to grow as a company. They enlisted the expertise of Wimbledon based office design and build company Peldon Rose to help with this growth.

It was decided that the second largest and tallest building to date at Chiswick Park would be the perfect space for their London HQ. Peldon Rose has previously completed a number of projects at Chiswick Park, including Baker Hughes, Starbucks and Tullow Oil.

But what makes this project different to any other previously designed and built — The fact that it is the largest design and build project in the UK and one of the largest in Europe at a massive 19,695 m^2.

Peldon Rose transformed the huge existing traditional call-center office set-up with angular clusters of adjustable sit-stand desks to create an environment where people can work creativity, collaboratively, and in a contemporary space.

Aker Solutions are thrilled with their new office design, and commented, "Peldon Rose worked with us to develop a high quality brief and delivered beyond our expectations."

One of Peldon Rose's core values is a "Client for life" and they've really lived this. Now the project has been completed, Peldon Rose appointed a dedicated member of staff that Aker Solutions can go to if they ever have any problems with their new office.

COLORS MATCH
色彩搭配

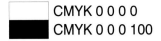
CMYK 0 0 0 0
CMYK 0 0 0 100

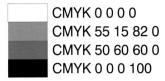
CMYK 0 0 0 0
CMYK 55 15 82 0
CMYK 50 60 60 0
CMYK 0 0 0 100

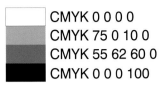
CMYK 0 0 0 0
CMYK 75 0 10 0
CMYK 55 62 60 0
CMYK 0 0 0 100

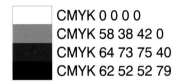
CMYK 0 0 0 0
CMYK 58 38 42 0
CMYK 64 73 75 40
CMYK 62 52 52 79

Peldon Rose 设计事务所为他们最新的阿克集团项目设计倾尽全力。阿克集团（石油和天然气供给公司）正在寻求扩张，努力扩展为可以吸引新一代工程师的空间，并且可以使他们成长，撑起一个公司。他们招募了总部位于温布尔顿的办公室设计的专业人员和 Peldon Rose 设计事务所，以帮助他们实现这一发展目标。

他们决定将奇斯威克园区迄今为止第二大和第二高的建筑作为伦敦总部的完美选择。Peldon Rose 设计事务所此前完成了多项位于奇斯维克园区的项目设计，包括贝克休斯公司、星巴克集团和塔洛石油。

但是这个项目又区别于以往所有的设计和建造项目，它是英国最大的设计建造项目，并且凭借 19695 平方米的巨大建筑面积，成为了欧洲最大的项目之一。

Peldon Rose 设计事务所改变了现有的巨大的传统型呼叫中心办公结构，用可调节的坐立式的长方形会议桌取而代之，为人们营造了一个可以在现代化的空间里，创造性且协作式工作的办公环境。

阿克集团对新办公室的设计极为满意，并评价说，"Peldon Rose 设计事务所与我们合作开发出了远远超出我们预期的高品质而简洁的设计方案。"

Peldon Rose 设计事务所的一个核心价值观是"终身为客户"，他们也确实做到了。目前该项目已经完成，Peldon Rose 设计事务所还任命了该项目的客户专员，专门解决阿克集团以后在新办公室使用过程中可能出现的问题。

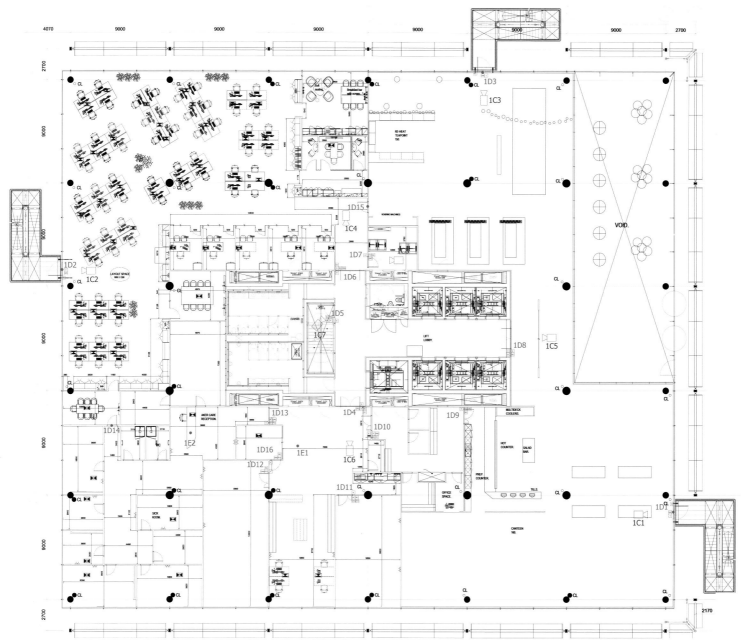

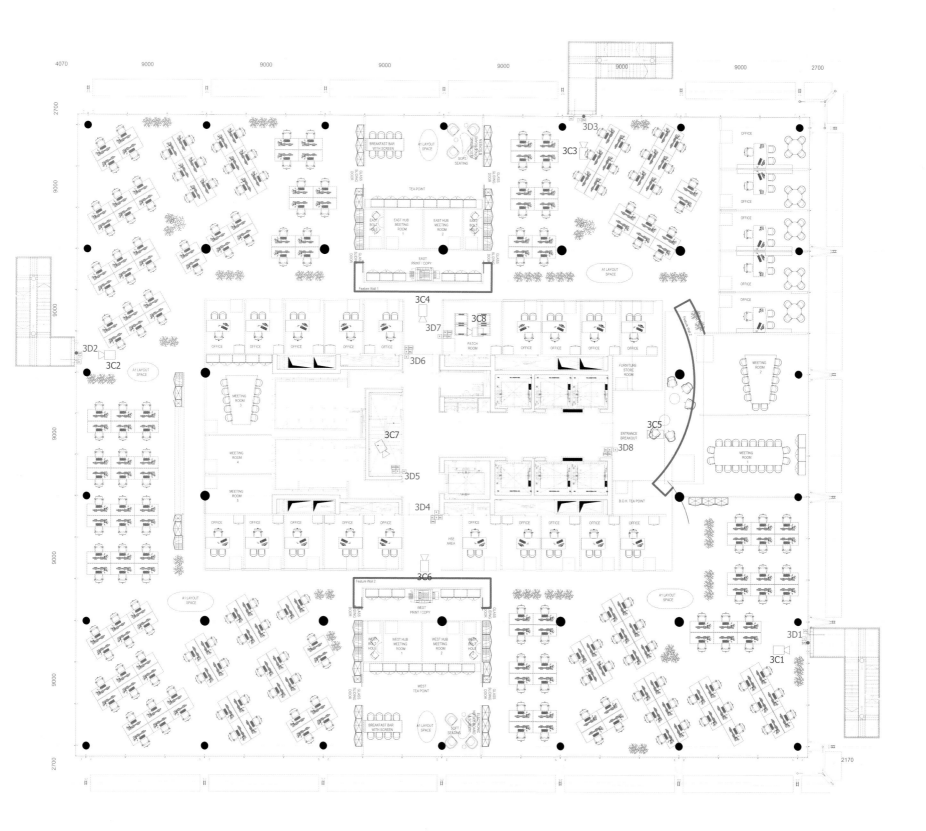

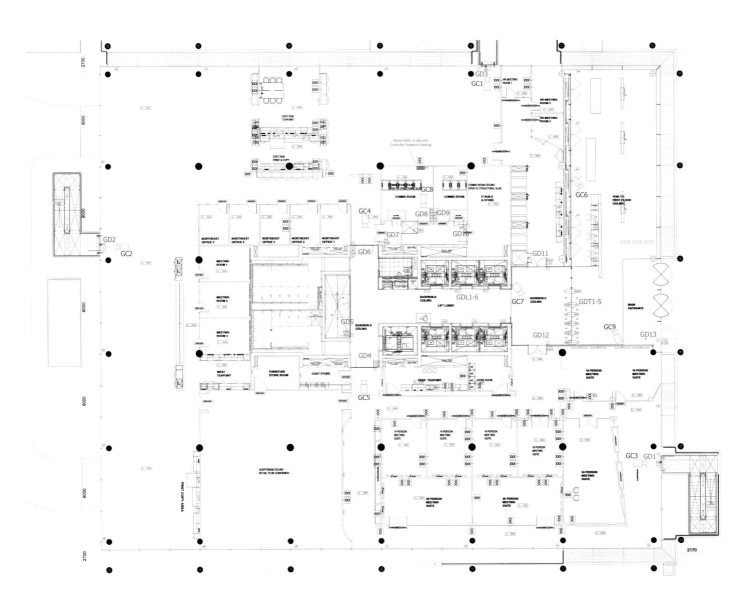

ALCATEL LUCENT
HEADQUARTER

阿尔卡特朗讯总部设计

*Designer*_DEGW Italia / *Project Manager*_Eva Birch (DEGW Italia) / *Technical Plant*_Roberto Cereda (L22)
*Works Supervisor*_Roberto Cereda (L22) / *Building Design*_Garretti Associati (Client Segro)
*Contractor*_ISG / *Client*_Alcatel Lucent, Alessandro Adamo (Leader), Marco Agazzi (Referent), Antonio Vitolo (Project Manager)
*Furniture*_Kinnarps, Steelcase, Sedus, Vitra, La Cividina, Pedrali, Luceplan / *Standard Furniture*_Unifor, Sedus, Elbo
*Custom Made Furniture*_Tekno Par and Underline / *Lighting*_Flos / *Moveable Walls*_Underline (Partition Lema)
*Plant*_L22 / *Location*_Vimercate (MB), Italy / *Area*_33,000 m² / *Photographer*_Dario Tettamanzi

"THE RESULT IS A COLORFUL, FUNCTIONAL, FLEXIBLE, BRIGHT WORK SPACE ABLE TO FOSTER BOTH COLLABORATION AND PRIVACY ISSUES, WELL ALIGNED WITH BUSINESS OBJECTIVES AND BRAND VISION.

COLORS MATCH
色彩搭配

CMYK 45 10 100 0
CMYK 42 50 65 0
CMYK 8 15 30 0
CMYK 80 30 40 0

CMYK 0 92 95 1
CMYK 8 15 30 0
CMYK 0 0 0 15

CMYK 0 0 0 0
CMYK 45 10 100 0
CMYK 20 35 45 0
CMYK 68 88 28 0

Alcatel Lucent needed a new headquarter: comfortable, technological and sustainable" says Alessandro Adamo DEGW Italia CEO " So we have supported our client during all the process, from occupier strategy to interior design, including 7,000 m² laboratories sustainability planning, where energy efficiency and architectural integration of plants criteria have been successfully applied."

The new headquarter hosts 1700 persons on 33,000 m² area in Segro Energy Park, near to former headquarter. The campus is composed of a extensive hall acting as a central hub connecting 5 multi-storey volumes with different functions (offices, labs and multimedia center).

The 2,300 m² hall is a double height space (4~7 meters) : fluid, open and articulated in special areas (reception and waiting area, informal and formal meeting, client area, a cafeteria).

Adjacent to the hall, it is located the Multimedia Communication Center, a 1,200 m² space hosting 150 seats auditorium with a foyer, a small 60 places amphitheater, and the university including meeting and training rooms.

Into multilevel buildings are located offices (16,000 m²) and laboratories (6,800 m²).

Office floor plan type optimizes building dimensions (22x80 meters) delivering open spaces, that stimulate interaction, along with small private offices, when concentration is needed.

The central hub is conceived as a sharing workstation and hosts touchdown areas, concentration rooms, copy areas and groups archives.

DEGW interior design featured a choice of natural materials such as wood, and neutral colors, white and grey, jazzed up by fluo color furniture.

The laboratories show technological language by using high-tech materials: metal, glass, plants.

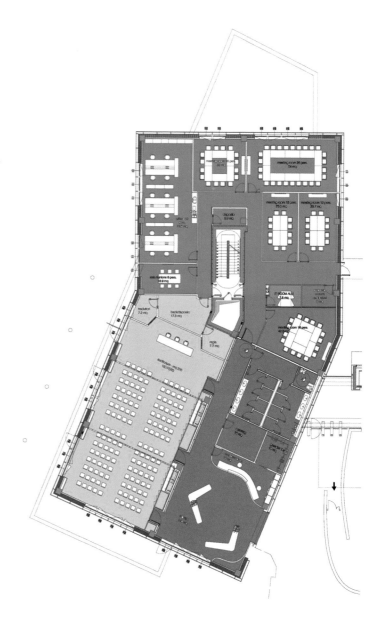

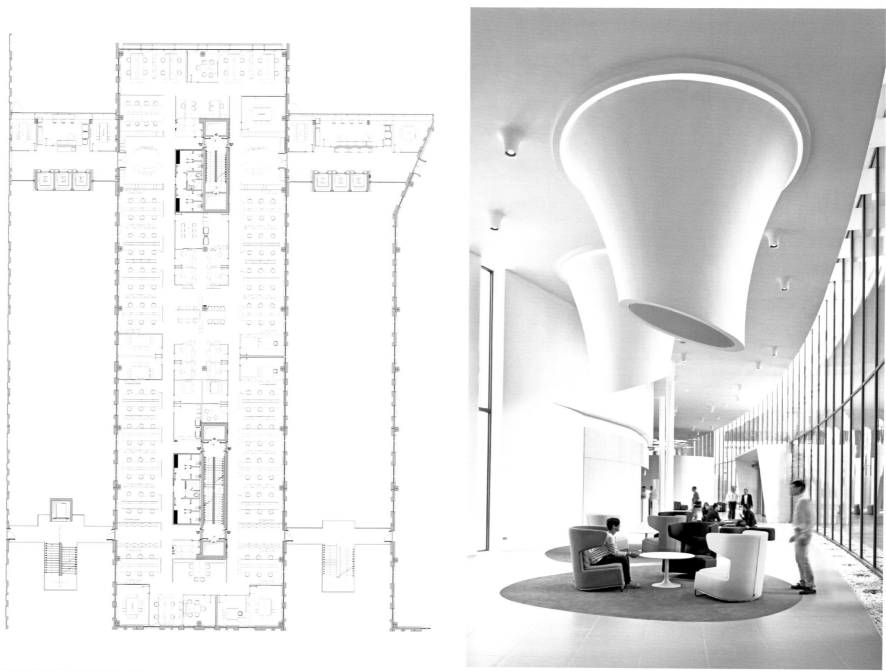

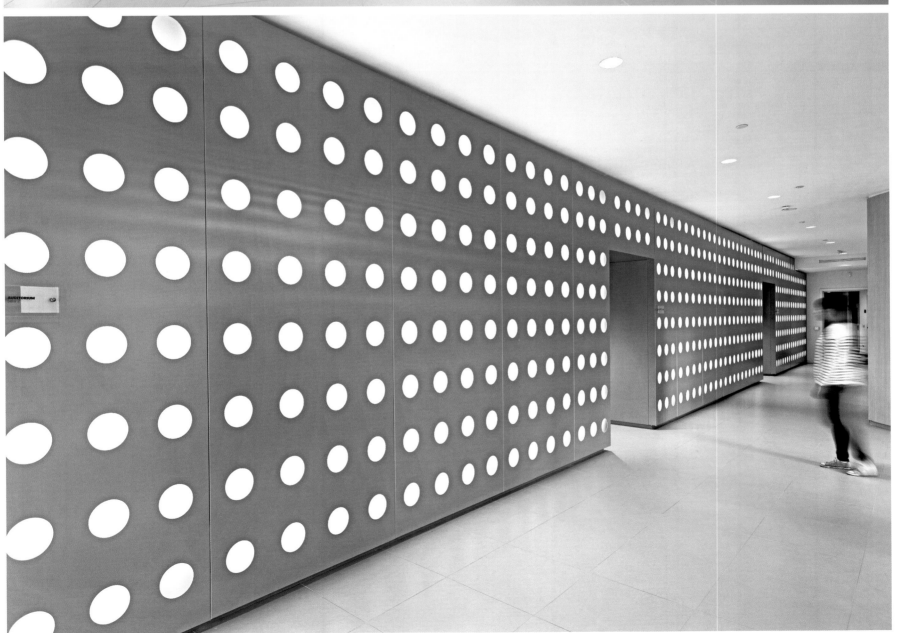

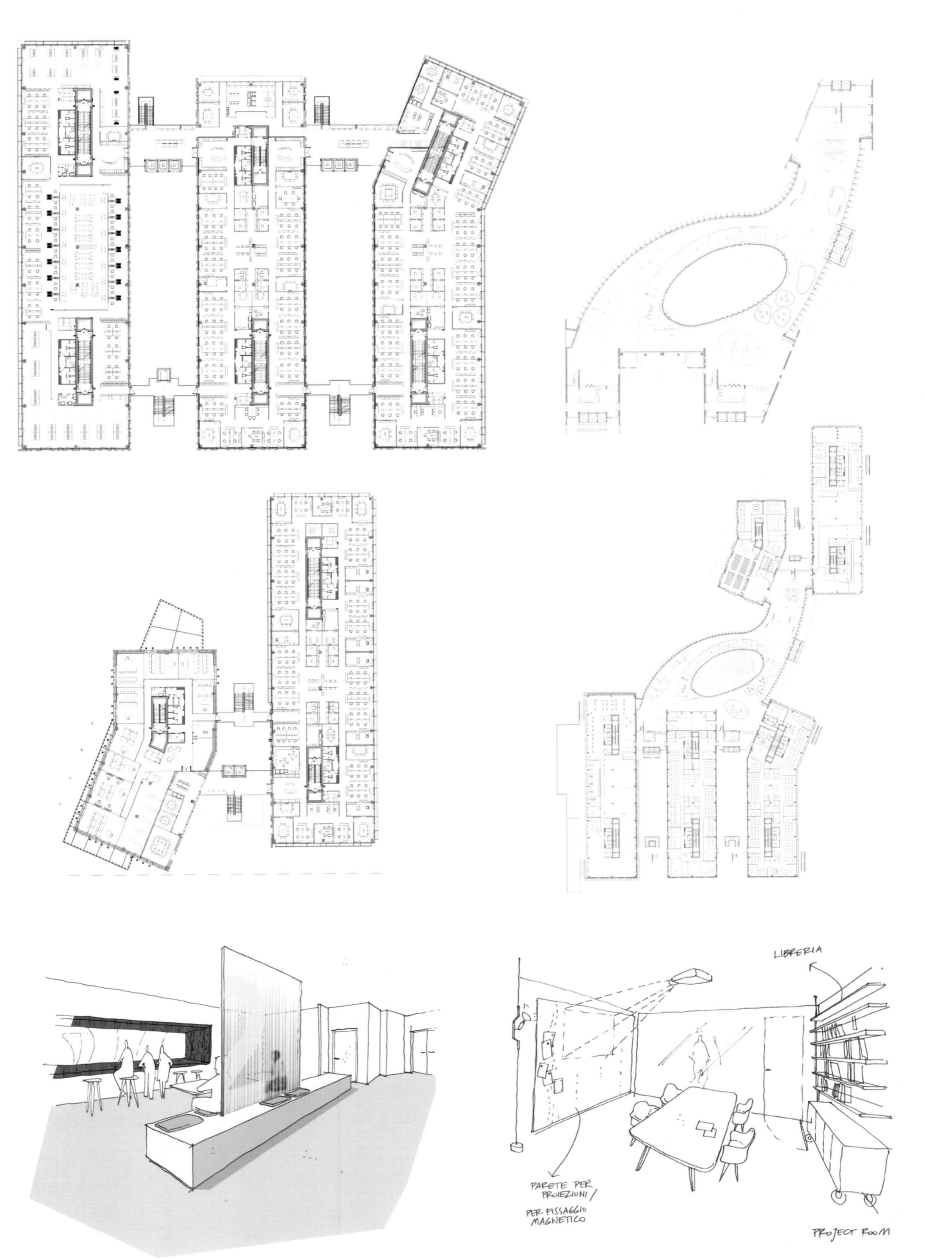

"本案目的是建造一个丰富多彩、功能齐全、灵活且明亮的工作空间，员工既能相互协作又能保持隐私，共同向着商业目标和品牌愿景前进。阿尔卡特朗讯需要一个新的总部：舒适、高科技并且保证可持续发展"，DEGW Italia 的首席执行官亚历山大·阿德莫说，"因此，我们全程支持我们的客户，从使用策略到室内设计，包括 7000 平方米实验室的可持续发展规划，在这个规划里，能源效率和工厂标准的建筑一体化已经得到成功的运用。"

新总部位于 Segro 能源园区，靠近前总部，占地 33000 平方米，能容纳 1700 人。园区包括一个起中心枢纽作用的大厅，连接着五层具有不同功能的区域（办公室、实验室和多媒体中心）。

该面积为 2300 平方米的大厅占据双层空间（4~7m 高）：流畅、开放、并且在特殊区域功能共享（接待处和候车区、非正式和正式会议室、客户区、餐厅）。

多媒体通信中心紧挨着大厅，面积 1200 平方米，包含一个可容纳 150 个座位的礼堂、门厅、一个能容纳 60 人的小型露天剧场和大学的会议室和培训室。

多层建筑里面有办公室（16000 平方米）和实验室（6800 平方米）。

办公室平面结构优化了建筑规模（22m×80m），并留出开放的空间用于互动交流，同时也设有用于员工集中精力工作的小型私人办公室。

中央枢纽被设想为一个共享的工作站，包括交流区、集中精力工作室、打印区和分组档案室。

DEGW 室内设计的特色是选用天然材料，如木材；并且使用中性的色调，例如，白色和灰色，并用荧光绿色调的家具使之活泼明朗。

实验室通过使用高科技材料金属、玻璃以及植物表达出一种科学技术的气息。

FLEETPARTNERS

FleetPartners 办公区

*Designer*_Gray Puksand / *Location*_Richmond, Melbourne, Australia
*Area*_3,600 m² / *Photographer*_Mark Duffus

THIS AUTOMOTIVE FINANCE PROVIDER IS THE CLEAR LEADER IN THEIR FIELD.

The workspace design reflects FleetPartner's market position and expresses their desire to retain that position.

The design captured the following concepts, drawing upon the inspiration of the automobile.

Concept: Reception zone — reflecting the aerodynamic curves and smooth lines.

Create: Workspace — The engine room, expose the workings, fixings and fastenings. No covers, do not hide the hard work.

Excellence: Breakout — Inspired by the luxury of success, warm and sumptuous materials.

The result is a space where the staff understand what the culture is, and also provides a place to inspire all who enter.

COLORS MATCH
色彩搭配

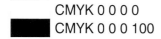
CMYK 0 0 0 0
CMYK 0 0 0 100

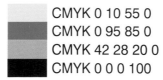
CMYK 0 10 55 0
CMYK 0 95 85 0
CMYK 42 28 20 0
CMYK 0 0 0 100

CMYK 0 32 65 0
CMYK 0 0 0 0
CMYK 0 0 0 100

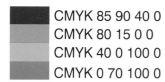
CMYK 85 90 40 0
CMYK 80 15 0 0
CMYK 40 0 100 0
CMYK 0 70 100 0

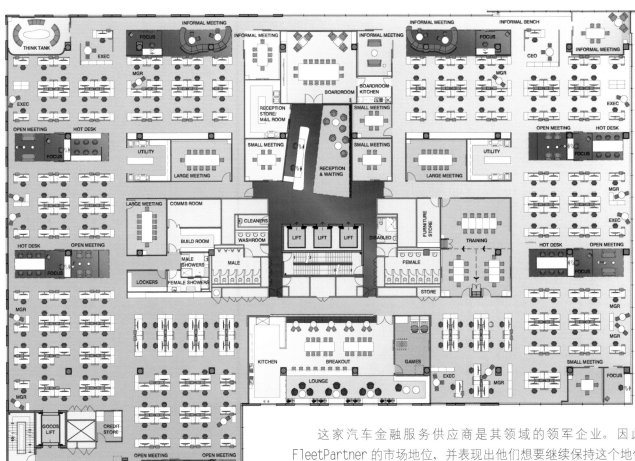

　　这家汽车金融服务供应商是其领域的领军企业。因此，工作区的设计展现了 FleetPartner 的市场地位，并表现出他们想要继续保持这个地位的愿望。

　　这个设计体现了以下理念，并从汽车中得到了设计灵感。

　　理念：接待区域——充分表现出空气动力曲线和平滑线的特点。

　　创新：工作区——发动机房，展示了操作方式、设备和连接件。无遮挡，丝毫不掩饰艰辛的工作。

　　卓越：突破——灵感源于成功后的享受和温暖、豪华的材料。最后创造出一个让员工理解文化内涵并能激发人们灵感的空间。

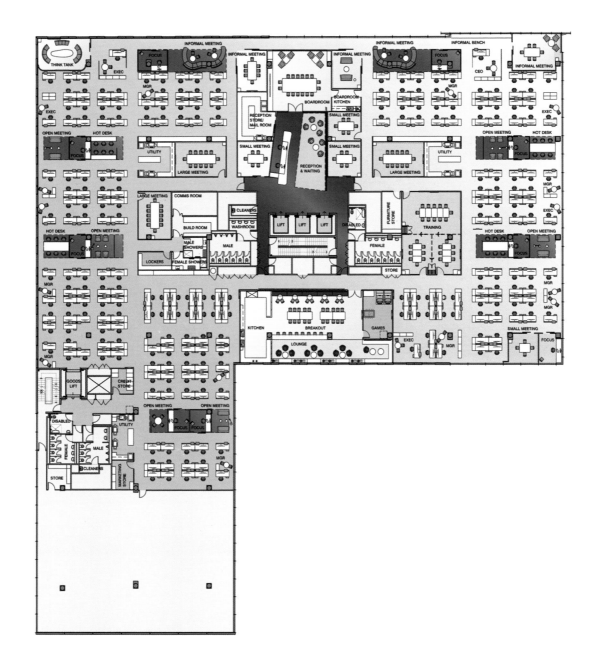

GOOGLE MADRID HQ

谷歌马德里总部

*Designer*_Jump Studios / *Project Manager*_Glenn Welland, Marta Alda Estables, Manuel Domenech, Paloma Guinea Merlot (Assistant), Artelia Spain (Site), Eva Jung (Facility) / *Project Executive*_Christian Hurzeler
*Director*_Simon Jordan (Managing), Pablo Sainz De Baranda (Managing), Miquel Castelvi (Managing), Shaun Fernandes (Creative), Juan Escabias (Project)
*Project Architect, Lisbon Office Head*_Laszlo Varga / *Local Architect*_Josefina Aldasoro
*Engineering, Architect of Record*_Deerns Spain / *Sustainability*_Indra Spain
*Leed Consultant*_Daniel Martín Hernández / *Contractor*_Construcía / *Client*_Google / *Photographer*_Daniel Malhão

THE EXTENSIVE FIT OUT AND REFURBISHMENT OF GOOGLE'S MADRID HQ SETS NEW STANDARDS IN OFFICE INTERIOR DESIGN ON THE IBERIAN PENINSULA.

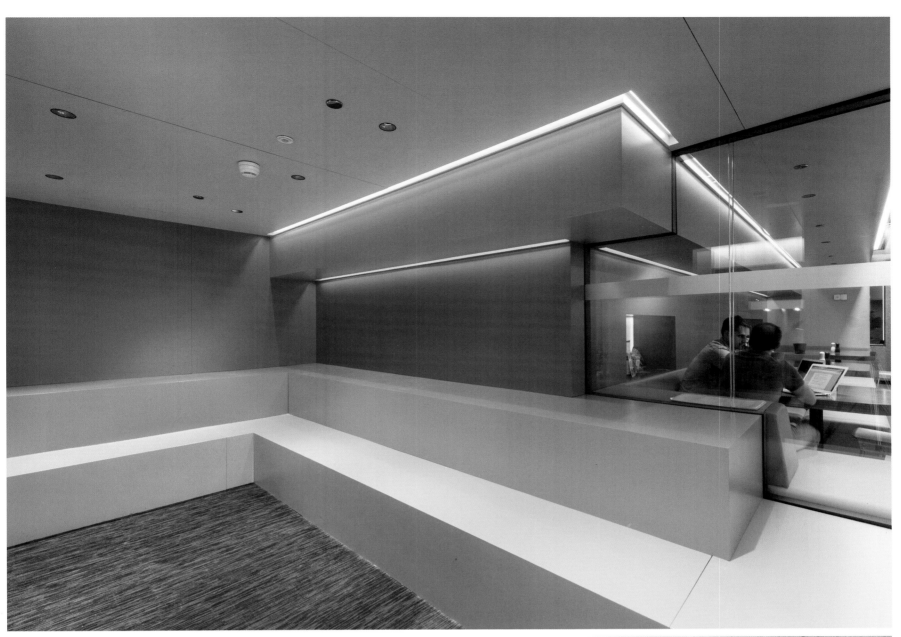

COLORS MATCH
色彩搭配

CMYK 0 0 0 0
CMYK 10 95 75 0
CMYK 63 58 65 25

CMYK 95 80 0 0
CMYK 95 50 0 0
CMYK 60 20 0 0
CMYK 0 0 0 30

CMYK 0 19 60 0
CMYK 10 95 70 0
CMYK 0 65 60 0
CMYK 0 0 0 60

CMYK 65 0 90 0
CMYK 53 58 78 0
CMYK 0 0 0 50
CMYK 10 0 0 80

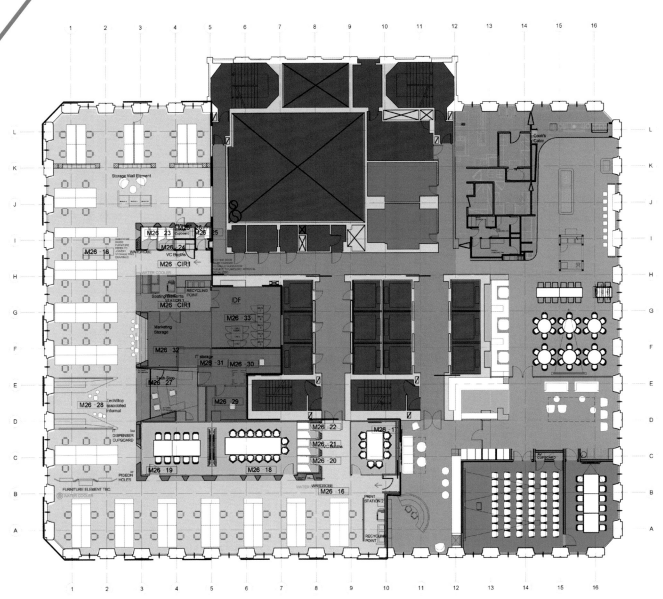

The lower of the two adjacent levels occupied by the client houses the main reception, lecture theatre, canteen and a multi-functional area with fully equipped kitchen catering for the entire office.

On the upper level can be found the bulk of the office space as well as more extensive breakout spaces with room for games, additional informal presentation areas, shower facilities, a massage room and hammock area.

The overall layout and arrangement of particular spaces and elements has been carefully considered and developed to suit the working style of the company in general while meeting the more exact needs and requirements of the local workforce.

The very specific acoustic requirements of the project for both the meeting rooms and the individual video conferencing cabins necessitated the careful selection of subcontractors and the very close co-ordination of all the teams involved to provide both robust and aesthetically pleasing solutions and details. The use of sustainable materials contributed to the project's LEED Gold rating.

The project was delivered in five separate phases, which allowed the offices to remain open throughout. It involved a high level of co-ordination and collaboration between the architectural, engineering and contracting teams – Jump Studios, Deerns and Construcía with strong project management from Artelia Spain.

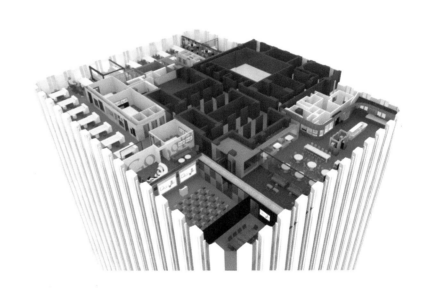

谷歌马德里总部大楼的装配整修为伊比利亚半岛的室内设计制定了新的标准。

下层两个毗邻的空间包括客户接待区、报告厅、餐厅和多功能区（内含设备齐全的厨房，可服务于全体办公室成员）。

上层空间包括宽阔的办公区和宽敞的游戏区、非正式汇报区、淋浴、按摩和休闲吊床区。

大楼的整体布局设置考虑到了特殊空间和元素的使用，并体现了公司的工作风格，满足了员工更多、更细致的工作需求。

承包商精心挑选了满足会议室和视频会议室对音质要求的材料，所有团队紧密合作并提供实用性与美学上的解决方案和处理细节。可持续材料的使用有助于这个项目实现绿色建筑最高等级认证。

整个项目包括五个独立阶段，这样可以使办公室继续运作。此项目涉及建筑团队、工程团队、承包商之间高水准的相互协调和合作，其中包括 Jump Studios, Deerns 和 Construcía 公司，以及具有较强工程管理能力的西班牙 Artelia 公司。

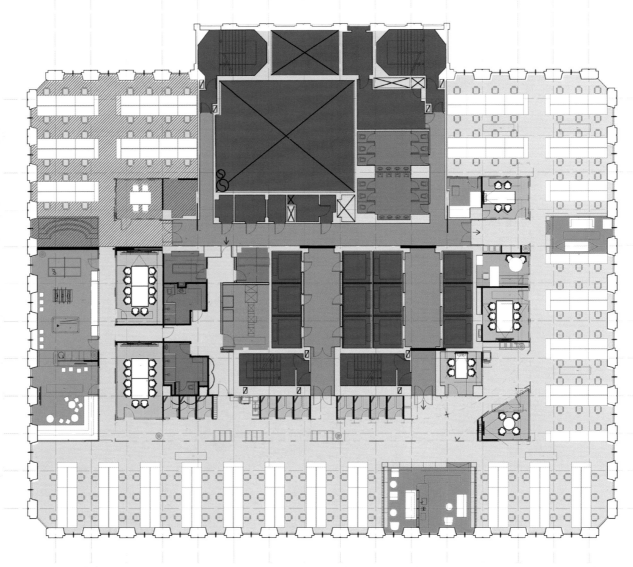

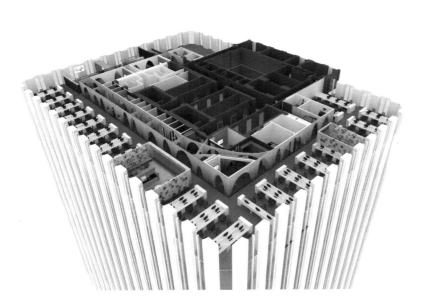

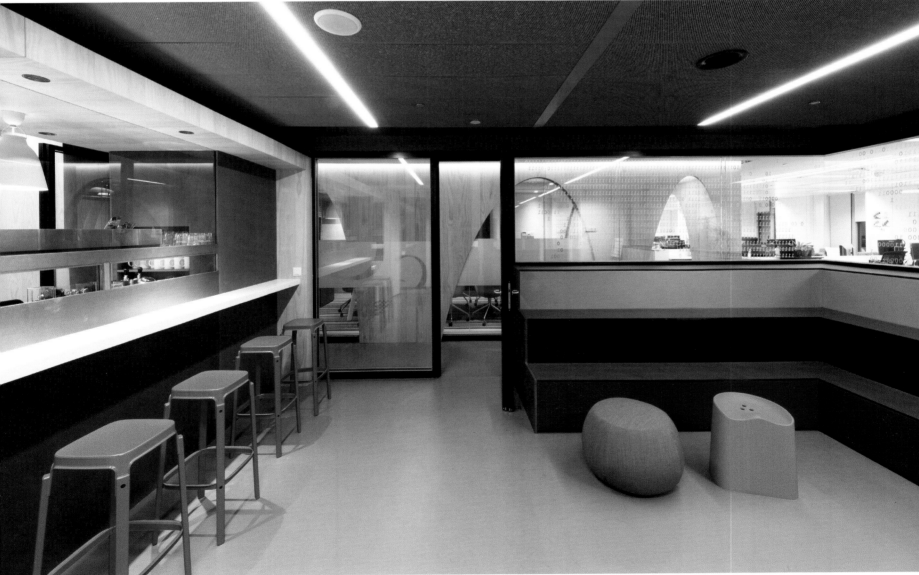

MASISA
OFFICE

Masisa 办公室

Designer_Arch. Guto Requena / Design Company_Estudio Guto Requena
Co-Designer_SP Estudio, Arch. Tatiana Sakurai / Development_Arch. Paulo de Camargo
Construction Architect_Ufficcio Architecture and Engineering / Lighting_Omega Light
Carpentry_Ronimar Carpentry / Furniture_Bertolini, Herman Muller, Dpot and Saccaro
Location_São Paulo, Brazil / Area_300 m² / Photographer_Fran Parente

WE CHOSE A COLOR CHART WITH NEUTRAL TONES SUCH AS GRAYS AND BEIGE, PUNCTUATED WITH GREEN ACCENTS.

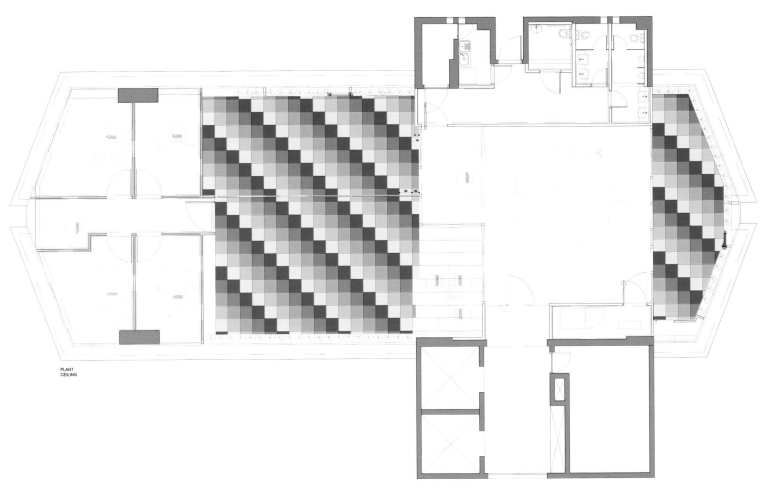

PLANT
CEILING

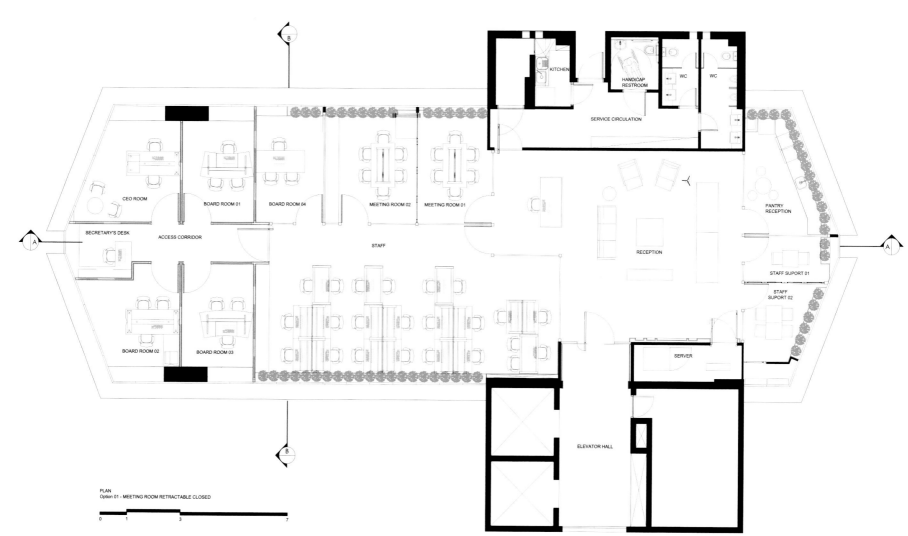

PLAN
Option 01 - MEETING ROOM RETRACTABLE CLOSED

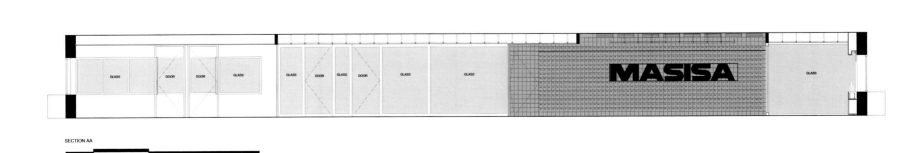

SECTION AA

0 1   3   7

我们挑选并使用中性颜色,如灰色和米黄色,并用绿色加以强调来进行办公室设计。每一步我们都采用能够通过家具展示巴西风格的方案,例如,在接待区,有来自国家级设计师 Fernando, Humberto Campana, Sérgio Rodrigues, Paulo Biacchi 和 Jader Almeida 设计的家具。我们将 masisa 品牌的木质板材应用于办公室、天花板四周、橱柜内部、浴室,以及服务区域,彰显了产品的多种用途。天花板图案部分,我们将木质装饰板以斜对角的方式布置,与周围的空间保持一致,并与所有工作区相融合。同样的设计也应用于地毯的剪裁上,地毯色调柔和,大小依据天花板的装饰而进行剪裁。灯光的设计强化了 Masisa 总部内部的设计,设计师选择了颇具动感的"z"字形灯具,并将其悬吊于天花板上。在接待区,我们设计了环绕式的分散灯光以营造热情舒适的氛围。我们还在接待区设计了霓虹灯,让入口处显得更醒目,更人性化。人们可以看到墙上的 masisa 公司的 logo,同时这面墙也是该公司木产品的展示墙,各种木板以艺术的布局安装在墙上,从视觉上让人联想起屏幕的像素点。所有呈现在屏幕上的装饰像素板都是可拆分的,建筑师和设计师可以自助拆分、设置,并在旁边的工作台上熟悉 Masisa 的产品。

一条绿色的植物带穿过办公区带来了巴西自然原色特点的设计,同时也提醒我们在工作中不要有官场式的冷漠,或者缺乏情感。所有 Masisa 圣保罗总部内部设计材料的选择都试图努力减少对环境的影响,充分利用当地资源和劳动力。

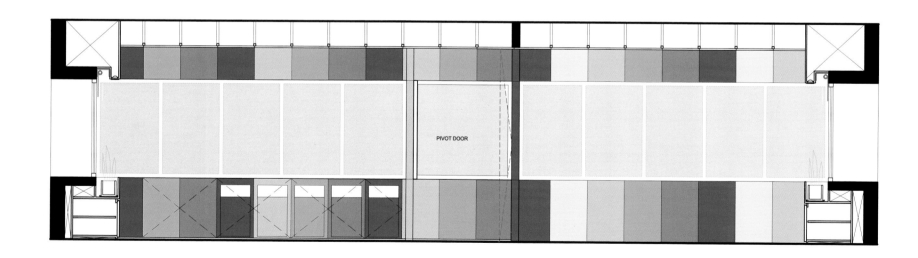

SECTION BB

OCA GROUP

OCA 集团

Designer_Denys & von Arend / Location_Madrid, Spain
Area_3,000 m² / Photographer_Victor Hugo

OCA IS A CERTIFICATION COMPANY RECENTLY RELOCATED IN AN OLD ENTIRELY RENEWED BUILDING IN MADRID AREA.

COLORS MATCH
色彩搭配

CMYK 0 0 0 0
CMYK 0 25 70 0
CMYK 0 0 0 100

CMYK 0 10 40 0
CMYK 0 65 65 0
CMYK 5 82 48 0
CMYK 70 54 0 0

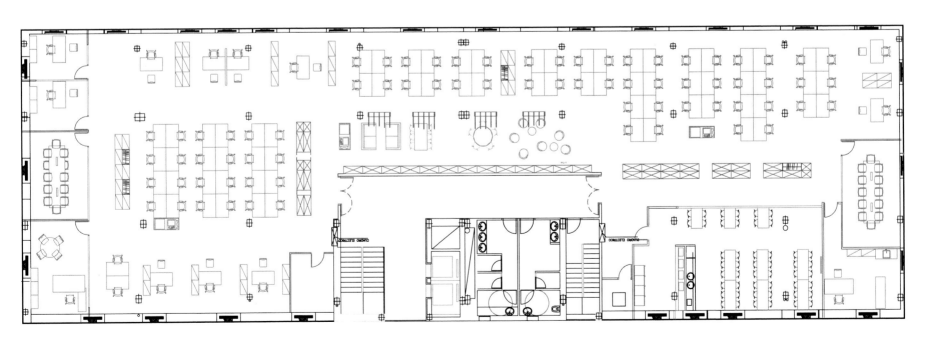

URALCHEM HEADQUARTERS

Uralchem 总部

Designer_Luis Pedra Silva e Maria Rita Pais / Design Company_Pedra Silva Arquitectos
Design Team_André Góis, Dina Castro, Hugo Ramos, Hugo Ferreira, João Alves, Paulo André, Ricardo Sousa
Contractor_VURAL / Carpentry & Mansory_BEC / Partition_OWD / Graphic Design_P06
Location_Moscow, Russia / Area_3,370 m² / Photographer_Fernando Guerra

THE BRIEF WAS TO CREATE A SPACE THAT WOULD ENABLE IDEAL WORKING CONDITIONS FOR STAFF WHILE ALSO REFLECTING THE COMPANY'S DYNAMIC, RELAXED AND YOUTHFUL SPIRIT.

COLORS MATCH
色彩搭配

CMYK 0 0 0 0
CMYK 0 0 0 75
CMYK 0 70 70 0
CMYK 15 28 48 0

CMYK 0 0 0 0
CMYK 30 5 78 0
CMYK 0 0 0 60

The project space occupies an entire floor of Imperia Tower, a skyscraper in Moscow City.

The ceiling is made up of white circular elements that form a continuous surface embracing the space and reflecting natural light. Artificial light is achieved by clicking in light discs according to the amount of light required around a particular space. You do the same by placing equivalent diameter air-conditioning vents that become fully integrated in the ceiling system. The result is a flexible system of interchangeable suspended disks, allowing for easy access to the upper infrastructure while minimizing the effect of a lower ceiling.

If we had used a regular office ceiling, the space would have felt cramped and claustrophobic, but instead it feels big and airy while still ticking all the boxes of the performance you need from a regular office ceiling. From initial prototypes in Innsbruck, Austria to actual production in Ankara, Turkey the result is a bespoke answer to the initial problem resulting in the main aesthetic element of this space.

As for the space itself, the office is arranged around the central service core of the building, working as a distribution nucleus between departments. The circulation route around this core is emphasized by a continuous wood surface that is randomly cut so it secretly hides storage units and doors.

Besides the aesthetic nature of the space, key spaces were provided for staff to promote well-being and productivity. Spaces such as silent rooms to improve concentration, small rooms for unplanned meetings, noise removing elements for the open space and a large coffee lounge where staff mingle and share their experiences.

The space is occasionally interrupted by "glass bubbles" that contrast in nature to surrounding circular references. These "bubbles" contain team leaders and noisy rooms within a sound proof environment. The pictograms add a splash of color in an otherwise calm and neutral environment.

该设计欲营造一个理想的工作空间,既能使员工享受较好的工作条件同时也能反应公司动态的、轻松而朝气蓬勃的精神面貌。此项目空间占据莫斯科一座摩天大厦(Imperia Tower)的一整层。

天花板是由白色的圆形元素构成的,白色圆形元素围绕着空间形成一个连续的表面并能反射自然光线。根据特定空间所需要的光量,点击光磁盘就可以实现人工照明,人工照明也可以通过放置当量直径空调通风口,并使其和吊顶系统完全结合来实现。这样就产生了一个可互换的灵活的悬挂磁盘系统,既便于使用上面的基础设施,又将对下面天花板的影响降到最小。

如果我们使用普通的办公室天花板,会感到空间拥挤、压抑,但是使用这样的吊顶就会使空间宽敞而通风,并可以通过敲击磁盘得到普通天花板的功能。从奥地利因斯布鲁克最初的雏形设计到土耳其安卡拉的实际成果,都为本案空间在最初美学元素处理上提供了参考。

对于空间本身,各办公室作为分配中心被安排在大楼的中心服务区周围,环绕中心服务区的走廊墙面上覆盖着木饰面,间或伴有看似随意的划痕,可以悄悄地遮盖储存单元和门。

除了空间的审美特性以外,还有一些空间用于提升员工的幸福感和工作效率。例如,一些安静的房间用来凝聚员工的注意力;小房间用于召开计划之外的会议;噪音消减设施被应用于开放空间;大的咖啡厅可供员工之间交流,分享自己的经历。

空间不时地出现"玻璃泡",从本质上和周围圆形的参照物形成鲜明的对比。这些"泡泡"包括团队领导室和隔音环境中较为嘈杂的房间。这种形象的设计为原本安静和平淡的工作环境增添了色彩。

WHEN WORK BEGAN WITH RED BULL ON THE REDESIGN OF THEIR NORTH AMERICAN OFFICE, IT QUICKLY BECAME CLEAR THAT THEY NEEDED TO ACQUIRE ADDITIONAL SPACE TO SUPPORT THEIR RAPID GROWTH.

RED BULL NORTH AMERICA

北美红牛

*Designer*_Wirt Design Group / *Location*_Los Angeles, CA, U.S.A
*Area*_3,623 m² / *Photographer*_YK Cheung Photography

COLORS MATCH
色彩搭配

CMYK 35 0 85 0
CMYK 40 100 100 0
CMYK 75 45 48 0

CMYK 0 0 0 0
CMYK 15 95 80 0
CMYK 78 62 0 0
CMYK 0 0 0 80

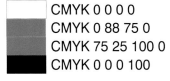
CMYK 0 0 0 0
CMYK 0 88 75 0
CMYK 75 25 100 0
CMYK 0 0 0 100

The client envisioned an energizing, egalitarian, interactive space — a sharp departure from their current quiet, cellular office — that would equally match their high-octane, risk-embracing brand. The resulting design direction was driven by this corporate decision to promote company culture through the work environment.

The design team was tasked to maintain as much of the existing architecture as possible to contain costs, while customizing the space to reflect the company's culture and create a tailor-made appearance. Planning focused on the following solutions to achieve these goals:

• Create an open, centralized community "hub" that would energize and connect the various departments within the large floor plate.

• Demolish the perimeter corner offices in order to accommodate alternative work/breakout areas and bring natural light and views into the interior space.

• Create a larger sense of space in the main work areas and reduce individual square footage though the use of benching systems with low dividing panels.

• Reconfigure some of the existing perimeter offices into additional conference and quiet rooms for privacy.

• Accommodate the balance of the program requirements within the remaining existing build-out, with a minimal amount of new construction.

In addition to the functionality of the space, conveying the company culture through integral branding was a key part of the design philosophy. Upon arrival, a dynamic, large-scale abstraction of the brand made of video monitors and red and gold reflective panels welcomes visitors and employees in the elevator lobby. Custom carbon fiber inlaid conference tables, commissioned graffiti art, motorcycles, surfboards and photos of sponsored athletes support brand identity and are carefully integrated into the design throughout the space.

在为北美红牛总部重新改造之初，我们非常清楚该公司需要额外的空间来支持它们快速的发展。

客户希望设计一个激励型、平等和交互式的空间——与他们当前安静、分格式办公室形成强烈反差——这同样与他们高能量而勇于冒险的品牌精神相匹配。公司期望通过优化工作环境来提升公司文化，这一意愿也决定了我们最终的设计方向。

北美红牛公司委派设计团队在为其设计个性化空间、反映公司文化、创建一个量身定做的外观的同时，尽可能地保持已有的建筑结构以控制成本。我们的解决方案如下：

在这座大楼内创建一个开放式、集中的团体活动中心，激励和团结各部门。

拆除周边角落的办公室，使其与其他工作区或扩展区相协调，并可以使室内空间有自然光线，有景观可供观赏。

在主要的工作区域创造更大的空间感，通过使用具有低矮分隔板的桌椅系统来减少个人使用面积。

将一些周边已有的办公室重新配置成额外的会议室和供私人使用的安静房间。

使项目需求和现存的扩建部分相平衡，尽量使新建部分最小化。

除了空间的功能性，通过整体的品牌来传递公司文化也是该设计理念的重要组成部分。一进公司，一个由视频显示器和红、金色反射面板组成的动感而庞大的品牌抽象作品便在前厅处迎接来访客人和公司员工。定制的碳素纤维镶嵌在会议桌里，涂鸦艺术、摩托车、冲浪板和赞助运动员的照片都服务于品牌标识，并被精心地融入到空间设计中。

CTAC'S – HERTOGENBOSCH

CTAC 斯海尔托亨博斯办公室

*Designer*_Eindhoven, Hans Maréchal, Bart Diederen, Marcel Visser / *Design Company*_M+R interior architecture
*Architect*_Joris Deur, Adam Smit (ZZDP architecten Amsterdam) / *Project Management*_Peak Development, Jaap van der Wijst
*Installaton Adviser*_Techniplan adviseurs / *Building Fysica*_LBP Sight / *Constructing Company*_Wessels Zeist
*Installation*_Bosman bedrijven / *Client*_CTAC 's-Hertogenbosch / *Fixed Furniture*_Odico interieurwwerken Deurne
*Furniture*_Desque Eindhoven, Arper, HAY, Satelliet; Stoffering: Kvadrat, Ohmann Leder / *Lighting*_Foscarini, Moooi
*Floor*_Desso / *Location*_'s-Hertogenbosch, the Netherlands / *Area*_3,750 m² / *Photographer*_Herman de Winter (studio de Winter)

CTAC SPECIALIZES IN IT AND BUSINESS CONSULTANCY. THE COMPANY OPERATES IN THE NETHERLANDS, GERMANY AND BELGIUM. ITS EMPLOYEES ARE OFTEN ON THE ROAD OR INVOLVED IN PROJECTS AT THE OFFICES OF DIFFERENT CLIENTS.

COLORS MATCH
色彩搭配

CMYK 0 0 0 0
CMYK 20 50 50 45
CMYK 38 75 18 0
CMYK 20 98 88 0

CMYK 30 0 95 0
CMYK 65 0 65 0
CMYK 0 0 0 100

CMYK 0 0 0 0
CMYK 0 85 95 0
CMYK 55 65 70 0

CMYK 0 0 0 0
CMYK 75 47 0 0
CMYK 55 65 70 0

The head office in 's-Hertogenbosch functions predominantly as a place where colleagues can meet. For the new office building on the highway A2, M+R designed an innovative working environment tailored to this need for informal and business meetings and flexible working.

Most of the interior is taken up by a grand café, an eating and working café, conference rooms and team rooms. The other rooms are intended as workspaces. These are designed and furnished in line with the length of stay and the needs of the staff, like the need for interaction or quiet, focused work. A total of five floors with approximately 750 square meters of floor space are being fitted out. When M+R got the mission of CTAC, counted the establishment in 's-Hertogenbosch some 550 employees. Initially the idea was to create as many fixed workstations.

M+R, however, advised the client to work flexible, because this would fit better in an innovative IT company as CTAC, where in addition, many employers work on location at the customers. M+R also developed the workplace concept, the new Office of CTAC now counts about 280 workplaces-most flexible-spread over a helpdesk, a front and a back office. M+R created in the Interior a clearly combination of the identity of CTAC and to show in which the company stands out. In establishing that identity keywords came up as Brabant (the area in the South of Holland), no-nonsense, hospitality, no one in suits, open and transparent, but introverted.

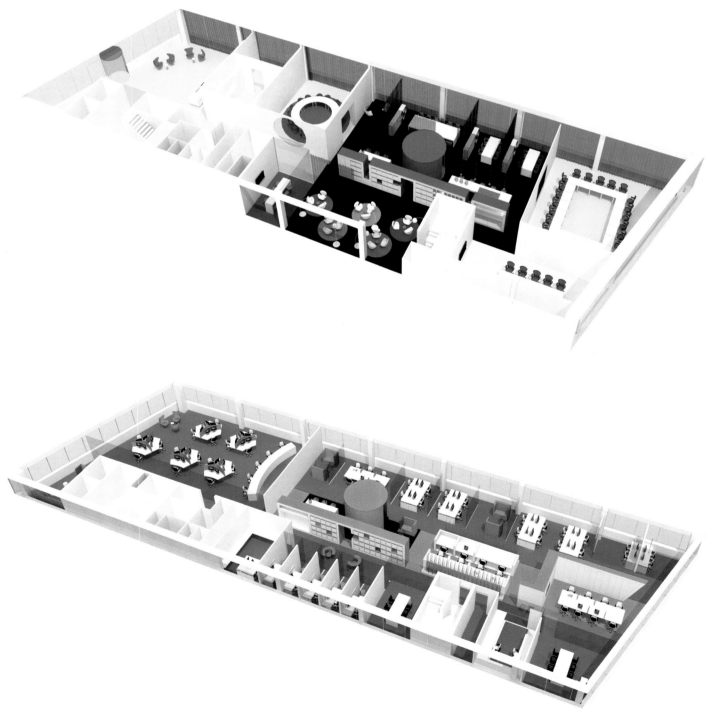

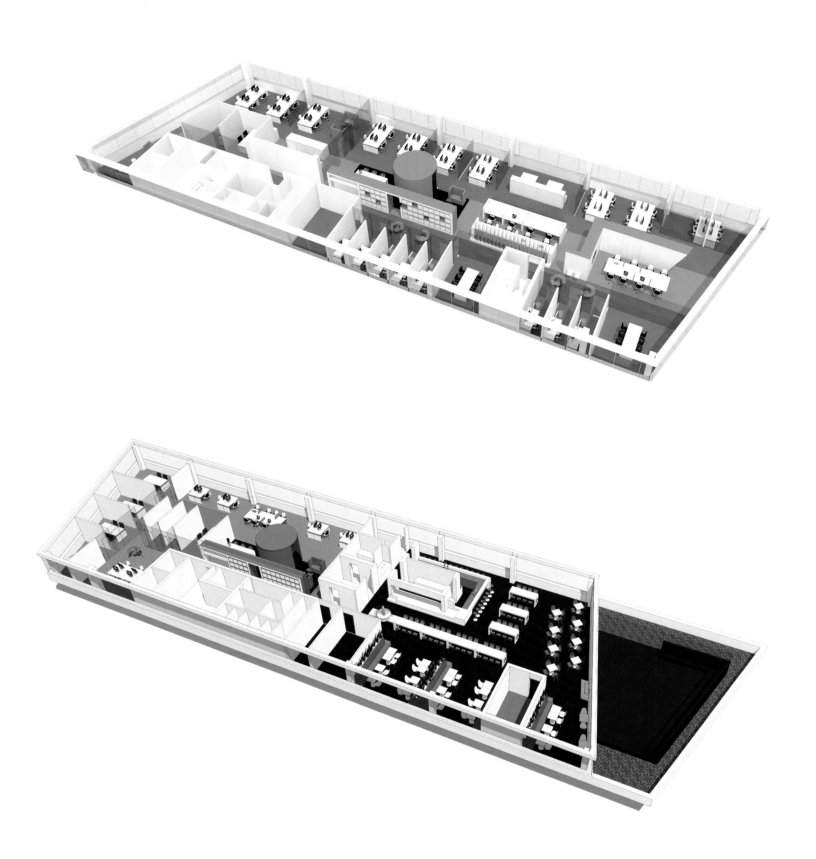

　　CTAC 公司专注于 IT 和商务咨询领域，该公司在荷兰、德国和比利时均设有分公司。员工往往不是忙碌在路上就是在不同客户的办事处参与项目。位于斯海尔托亨博斯的总部主要是作为同事相聚的地方。对于在高速公路 A2 处的新办公大楼，M + R 设计了具有创新性的工作环境，专门满足这种非正式会议、商务会议和弹性工作的需求。

　　办公楼内部大部分被豪华咖啡厅（可供用餐和工作的餐厅）、会议室和团队办公室所占据，其他房间均为工作区。这些设计和装饰符合员工的工作时间和需求（员工之间相互沟通的需要或安静专注于工作的需要）。办公楼共 5 层，约 750 平方米的建筑面积正在布置中。M + R 拿到 CTAC 的项目时，斯海尔托亨博斯 550 名员工都被考虑在内。最初的想法是创造尽可能多的固定工作站。

　　但是，M + R 建议客户灵活工作，因为这将更好地适应像 CTAC 这样的创新型 IT 公司；除此之外，很多雇主会在现场与客户商谈合作。M + R 还发展了工作场所的概念，CTAC 的新办公室算起来有 280 个工作场所（大多数是灵活的）分布在咨询台和管理及事务部门的周围。M + R 在内部设计时结合了 CTAC 的地位，并且体现了公司的与众不同。为了彰显这种地位，发掘了一系列关键词，如布拉班特（荷兰南部的区域）、严肃、热情、随意、开放、透明但含蓄。

OFFICE SPACE IN TOWN – @WATERLOO

Office Space in Town – @ 滑铁卢

*Designer*_Peldon Rose / *Builder*_Peldon Rose / *Build Duration*_26 Weeks
*Location*_Southwark, London, UK / *Area*_3,840 m² / *Photographer*_Matthew Beedlev

THIS WAS THE THIRD PROJECT THAT PELDON ROSE HAD DESIGNED AND BUILT FOR OFFICE SPACE IN TOWN.

COLORS MATCH
色彩搭配

CMYK 0 0 0 0
CMYK 58 15 0 0
CMYK 0 0 0 90

CMYK 0 0 0 0
CMYK 15 95 90 0
CMYK 0 0 0 100

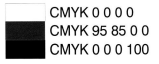

CMYK 0 0 0 0
CMYK 95 85 0 0
CMYK 0 0 0 100

CMYK 0 0 0 0
CMYK 15 95 90 0
CMYK 75 35 85 0
CMYK 0 0 0 100

And this would be the project to really push the boundaries for office design. Office Space in Town offers serviced offices, conference rooms and virtual offices throughout the UK.

The design concept came from taking a walk around Southwark, London – which was in fact the epicentre of hat making back in the day. However, the glue that was used to bind together the material to make the hats contained mercury. The mercury affected the minds of the craftsmen which gave the saying, "Mad as a Hatter". And so, the seed was sown, and an Alice in Wonderland inspired office was created!

Features included; suspended teapots instead of the usual lights, rosebushes that look like they've been freshly painted by the Queen's men, an overgrown, Alice who's disappearing through the ceiling. And of course, a giant white rabbit! Peldon Rose left no stone unturned with this Alice's Adventures in Wonderland themed design.

The six themed meeting rooms vary from stripped-back design – with Alice peering through the lit-up ceiling – to the outright whacky, like in the Mad Hatter room, where there's distressed wallpaper, grass on the floor and ceilings, and mismatched furniture.

Peldon Rose broke with convention and went with the 61:39 office-to-communal space ratio (compared to the usual 84:16) to encourage collaboration, creativity and productivity. And it's attracted more than just the creative start-ups. Even traditional businesses want to be in this space, and just four months in, Office Space in Town were at almost 100% capacity.

这是 Peldon Rose 设计事务所为 Office Space in Town 设计并建造的第三个项目，并且，此项目将使办公室设计更上一个台阶。Office Space in Town 在全英国范围内提供服务式办公室、会议室和虚拟办公室。

这一设计理念来源于在伦敦萨瑟克区的散步。萨瑟克区曾经是制帽中心，但是在制作帽子过程中，粘连材料所用的胶水中含有水银，这影响了工匠的智力，使工匠看起来都"疯疯癫癫"的。所以灵感来了，受爱丽丝梦游仙境启发的办公室设计应运而生。

该设计的特点有：悬吊式茶壶代替普通灯饰；仿佛是女王的手下刚刚画好的玫瑰丛；将要从天花板消失的巨人爱丽丝；当然，还有一个巨大的白兔子。Peldon Rose 为这个以爱丽丝梦游仙境为主题的设计倾尽全力。

六个主题会议室的设计各有不同：从通过点亮的天花板偷看的爱丽丝到彻头彻尾的怪诞，例如，在疯帽匠会议室，这里有受损的墙纸，地板和天花板上生长的草以及不搭调的家具。

Peldon Rose 打破陈规，使办公室与共用空间比率以 61:39 搭配（与通常的 84:16 相比较）来促进合作、激发创造力并提高生产力。这里不仅能吸引有创造力的新兴企业，甚至传统企业也为之所动，仅仅四个月时间，Office Space in Town 的"入住率"已达到 100%。

SITEGROUND

SiteGround

Designer_Cache atelie / Client_SiteGround / Location_Stara Zagora, Bulgaria Area_500 m² / Photographer_Minko Minev

CACHE ATELIE HAVE DESIGNED A 500-SQUARE-METER NEW OFFICE FOR SITEGROUND – AN IT COMPANY.

The office space is located in the city of Stara Zagora, Bulgaria and consists of two work space, conference room, meeting room and playroom. The main concept is developed around company's politics of offering handcrafted web hosting. Thus the main idea of paper planes emerges and is developed by Cache atelie into several variations of playful plane details. The eclectic mixture of vivid colors and playful wall decals, raw concrete ceiling, exposed infrastructural equipment and natural wood finishing creates unexpected yet harmonious space.

The enormous wooden doors with their specific detail act like an acoustic barrier and leave the great feeling of passing from one space to another. At the same time glass walls allow the light to penetrate all separate spaces and create muted feeling throughout the whole working space.

COLORS MATCH
色彩搭配

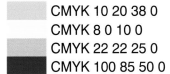

CMYK 10 20 38 0
CMYK 8 0 10 0
CMYK 22 22 25 0
CMYK 100 85 50 0

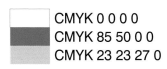

CMYK 0 0 0 0
CMYK 85 50 0 0
CMYK 23 23 27 0

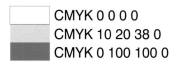

CMYK 0 0 0 0
CMYK 10 20 38 0
CMYK 0 100 100 0

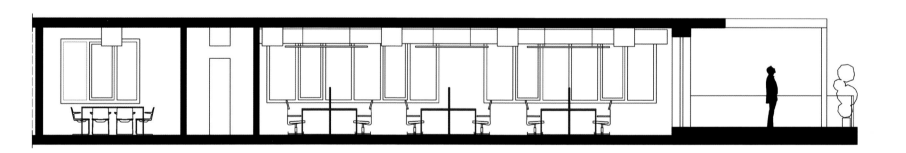

Cache atelie 设计团队为信息技术公司 SiteGround 设计了面积为 500 平方米的新办公室。该办公室位于保加利亚的旧扎戈拉市，有两个工作区、会议室、会客室和娱乐室。主要设计理念围绕着公司提供的量身定做的网络托管服务。因此，纸飞机的想法就应运而生了，并且 Cache atelie 设计团队把飞机细节做了很多有趣的改变。生动的色彩和好玩的壁纸的混合搭配，未处理的混凝土天花板，暴露在外的基础设施设备和天然的木材涂装营造出一个意想不到但十分和谐的空间。

巨大的木门像一个隔音墙一样，给人一种从一个空间进入到另一个空间的强烈感觉。同时，光线透过玻璃墙穿过所有的分隔空间，为整个工作空间营造了一种柔和的感觉。

URBIS SYDNEY

悉尼乌尔比斯

*Designer*_Gray Puksand / *Location*_Sydney CBD, NSW, Australia
*Area*_1,800 m² / *Photographer*_Brent Winstone

GRAY PUKSAND'S ENGAGEMENT COMMENCED WITH A FULL ASPIRATIONAL WORKSHOP TO BETTER UNDERSTAND THE URBIS CULTURE.

The workshops bring together a cross section of staff to review, challenge and propose new workspace opportunities, but also providing a conduit for the new information to be taken back to the entire group.

What arose was the desire for better communication, collaboration and a sense of fun. The result is organic and fluid forms of the floor, wall and ceiling surface to help navigate through the space and provide the much desired connection between clients, staff, internal teams and management.

Gray Puksand 设计公司的设计从梦寐以求的工作室开始，可以让人们更好地了解乌尔比斯的文化。这个工作室的设计汇集了不同部门的员工，他们回顾以往的工作空间设计并对其提出质疑，在此基础上提议新的工作空间设计，并将信息反馈给全体工作人员。

员工渴望更好地交流、合作以及工作的趣味性，因此，最终的设计将地面、墙壁和天花有机结合，连成一体，更易于员工在空间中穿行，也营造出客户、员工、内部团队和管理者渴望已久的交流感。

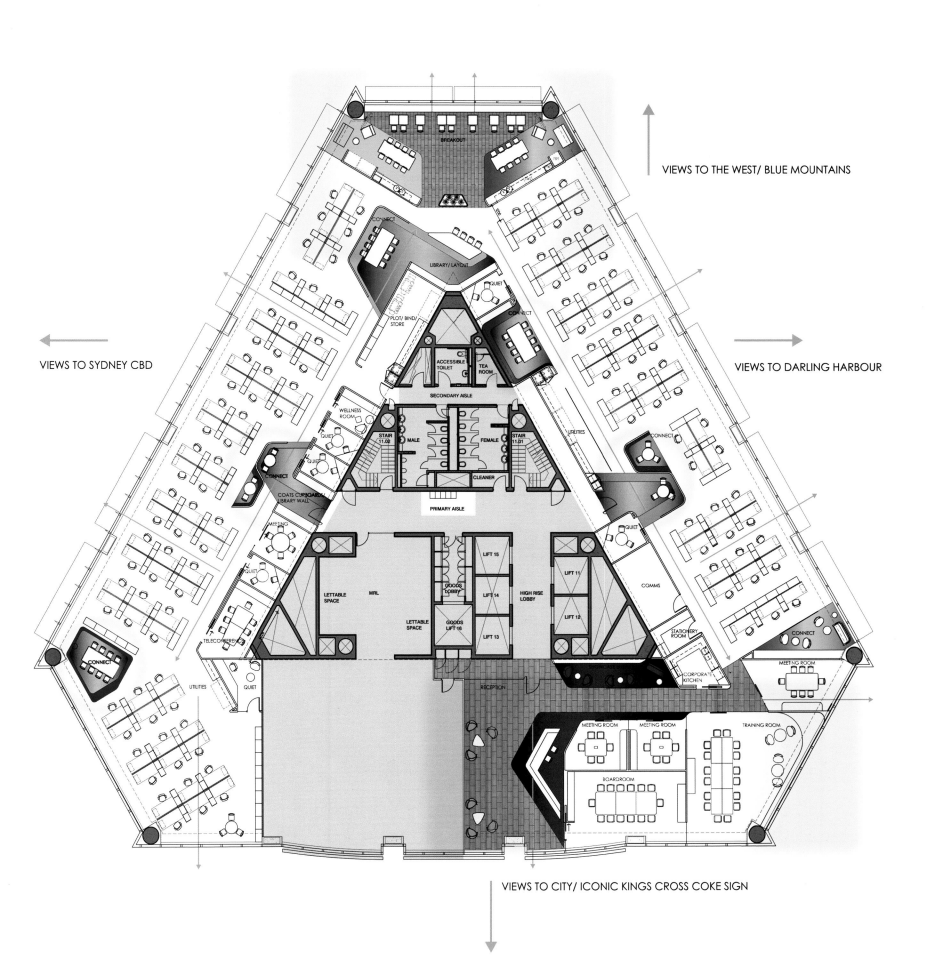

COLORS MATCH
色彩搭配

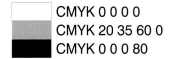
CMYK 0 0 0 0
CMYK 20 35 60 0
CMYK 0 0 0 80

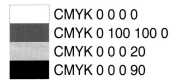
CMYK 0 0 0 0
CMYK 0 100 100 0
CMYK 0 0 0 20
CMYK 0 0 0 90

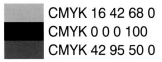
CMYK 16 42 68 0
CMYK 0 0 0 100
CMYK 42 95 50 0

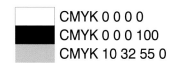
CMYK 0 0 0 0
CMYK 0 0 0 100
CMYK 10 32 55 0

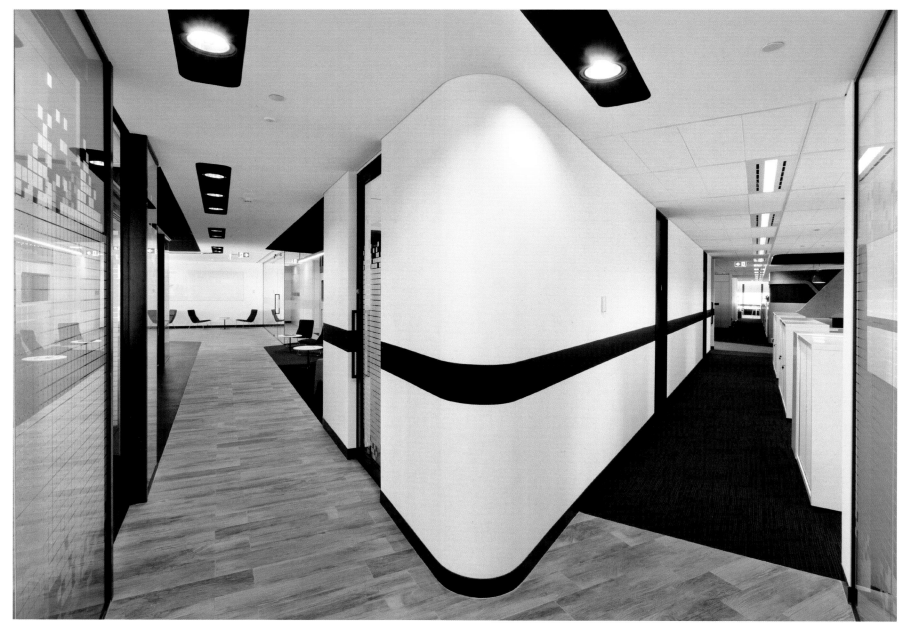

DREES & SOMMER
STUTTGART

Drees & Sommer 斯图加特办公室

Designer_Ippolito Fleitz Group – Identity Architects
Design Team_Peter Ippolito, Gunter Fleitz, Tilla Goldberg, Gideon Schröder, Roger Gasperlin,
Tim Lessmann, Trung Ha, Timo Flott, Tanja Ziegler
Lighting Design_Lichtwerke Köln, Stefan Hofmann / Client_Drees & Sommer AG
Location_Stuttgart, Germany / Area_2,775 m² / Photographer_Zooey Braun

AS AN INTERNATIONAL CORPORATION, DREES & SOMMER PROVIDES SUPPORT IN ALL ASPECTS OF PROPERTY FOR PUBLIC AND PRIVATE BUILDING OWNERS AS WELL AS INVESTORS.

Our range of services includes development consultancy, project management, engineering, infrastructure consultancy and strategic process consultancy.

The new office world features flowing spaces zoned by freely positioned island retreats and waist-height functional elements for printers and fax machines. The offices – separate but open to view – are perceived as part of the overall space.

The assistants' working areas are located centrally at the heart of the space and can be allocated to the various teams. The semi-circular table shape underscores their special function in the room.

The marketplace creates a central communication hub in the building that, like the in-house cafeteria, provides a space in which to take breaks, as well as being a setting for informal discussions and meetings with customers. An elongated compartment contains three booths, each seating up to four, for private conversations. In front of this is a long table for larger team meetings or sociable breaks.

The continuous wall-to-wall carpeting and suspended perforated ceiling grid integrate sound-absorbing surfaces into the building. In the more frequented circulation areas around the core, all the cabinet fronts and wall cladding are made of perforated acoustic boards.

In addition, fabric-covered panels between the individual workstations absorb noise emissions right at the source. Deep-pile carpeting ensures pleasant acoustics in meeting rooms. In the marketplace – the most intensively used communication zone – a highly absorbent ceiling of wooden slats complements and further improves the atmosphere and acoustics.

With the transformation of this office building, Drees & Sommer completes its own transformation towards a new working philosophy. The move away from a fixed workplace to a "non-territorial office" is fitting for a contemporary working environment in which flexible working hours and attendance times, as well as fluctuating team sizes, have become commonplace. This conversion of a 20-year-old building is a good example of the challenges facing countless properties due to the needs of a communicative working environment, today and tomorrow.

COLORS MATCH
色彩搭配

CMYK 0 0 0 0
CMYK 0 25 50 0
CMYK 0 0 0 50

CMYK 0 0 0 0
CMYK 50 45 50 10
CMYK 0 0 0 100

CMYK 0 0 0 0
CMYK 50 45 50 10
CMYK 0 0 0 100
CMYK 5 0 90 0

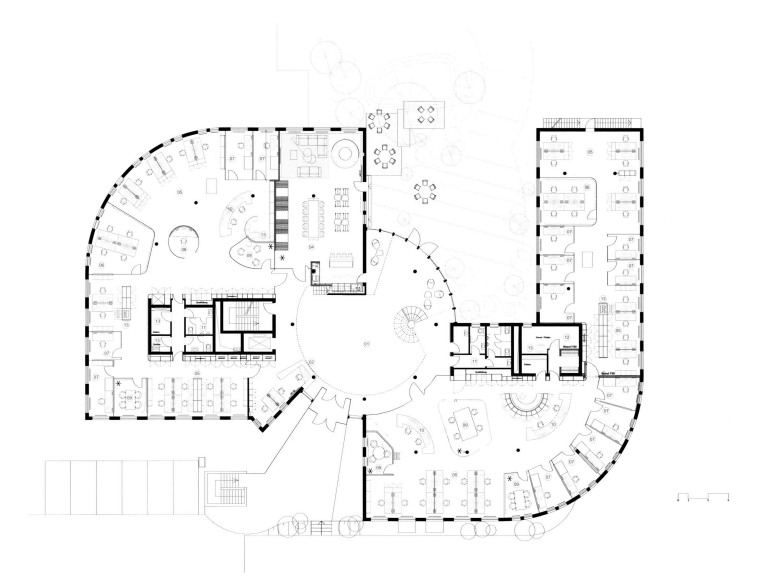

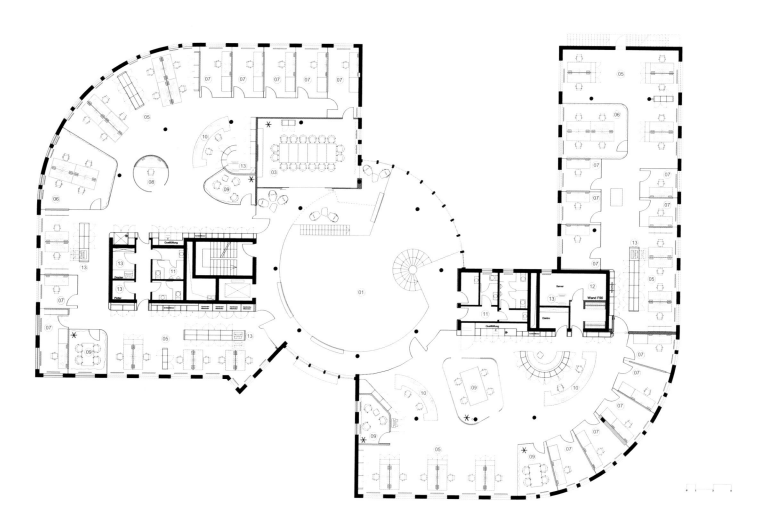

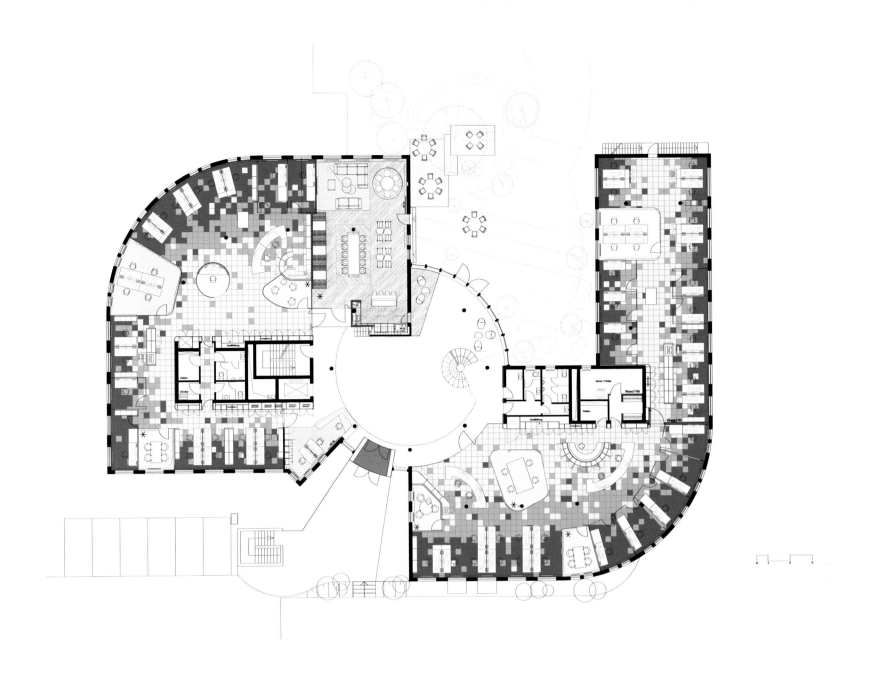
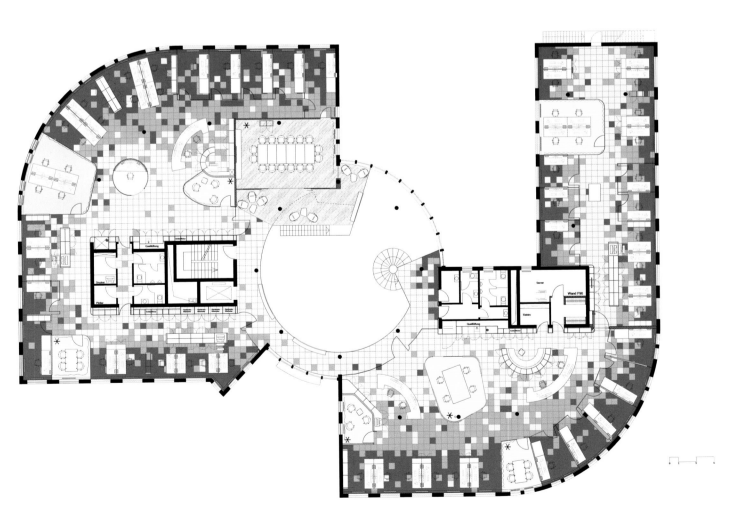

作为一家国际公司，Drees & Sommer 公司向公共和私人建筑业主以及投资者提供全方面的资产支持。我们的服务范围包括发展咨询、项目管理、工程、基础设施咨询和战略过程咨询。

新的办公世界以流动空间为特色，通过自由地设置小岛休息区、随意摆放高度及腰的打印机和传真机来划分空间。办公室彼此分开却有着开放的视野，是整体空间的一部分。

助理的工作区域位于办公空间的核心位置，可以通往不同的团队。半圆形的办公桌凸显其特殊功能。

卖场在建筑里创造了一个中央交流枢纽，像一个内部餐厅，为员工提供了休息的空间，并可以作为非正式讨论和会见客户的场所。细长的隔间包含三个小间，每个小间最多能坐四个人，用于私人聊天。在这前面是一个较长的会议桌，用于团队会议或休息时的沟通交流。

覆盖墙面的地毯和悬吊的多孔天花板网格将隔音表面引入建筑内部。在围绕核心区的较常用的沟通领域，所有的陈列柜前部和墙面覆盖层均采用穿孔吸声板。

此外，在一开始，工作站之间的织物覆盖面板就起到了隔音效果。厚实的地毯确保了会议室里的声音效果能够让人满意。卖场是最广泛利用的交流沟通区域，高度隔音的木质天花板进一步改善了环境并提升了声音的效果。

随着这幢办公楼的改造，Drees & Sommer 公司完成了其新的工作理念的转变。从固定的工作场所转向"无界限办公室"，更为符合现代的工作环境。现代工作环境允许灵活的工作时间和出席时间，也有着不断变化的团队规模，这已成为司空见惯的事。这个具有 20 年历史的建筑在工作环境上的改变为无数个现在或者将来希望实现互动、沟通式工作氛围的企业树立了榜样。

HSB

HSB

*Designer*_Peter Sahlin / *Design Company*_pS Arkitektur / *Project Architect*_Beata Denton
*Lighting Design*_Beata Denton / *Assisting Architect*_Martina Eliasson, Thérèse Svalling, Emilie Westergaard Folkersen
*Location*_Stockholm, Sweden / *Area*_9,000 m² / *Photographer*_Jason Strong Photography

HSB IS SWEDEN´S LARGEST HOUSING COOPERATION AND OWNED BY ITS MEMBERS. IT´S STOCKHOLM OFFICE HAS JUST UNDERGONE A COMPLETE RENOVATION IN ORDER TO ALLOW FOR OPENNESS AND ACCESSABILITY. ALL OF THIS APPLYING TO THE ENVIRONMENTAL CLASSIFICATION CALLED LEED "MILJÖBYGGNAD SILVER".

pS has been the interior architect and space planner. Much effort has gone into creating an ergonomic and modern office in terms of acoustics and lighting. This in combination with new technology has made the change from cellular offices to open workspace a pleasant experience.

Social interaction and energy has been the keyword and the theme is "Welcome home"! The reception, the so called "Living shop" and the inner courtyard all merge together on the ground floor, allowing for staff and guests to mix and mingle informally. The interior design is comfortable and colorful, contrasting efficiently against the original 40´ies.

Intarsia wall and pater noster lifts. The office spaces takes it´s inspiration from the city block. Each block consists of a number of desks and in the center there are "squares" and meeting points such as lounge furniture, hotdesks and telephone booths.

Some 420 people work in the building. The top floor has an amazing view over the city roof tops and presents a dozen or so meeting rooms for external meetings. Relaxing lounges and a creative space named "Think Tank" completes this welcoming office.

COLORS MATCH
色彩搭配

CMYK 0 0 0 0
CMYK 15 0 100 0
CMYK 5 25 70 0
CMYK 0 0 0 30

CMYK 0 0 0 0
CMYK 28 35 50 0
CMYK 40 85 0 0

CMYK 0 0 0 0
CMYK 0 83 24 0

CMYK 0 0 0 0
CMYK 0 0 0 0
CMYK 0 0 0 0

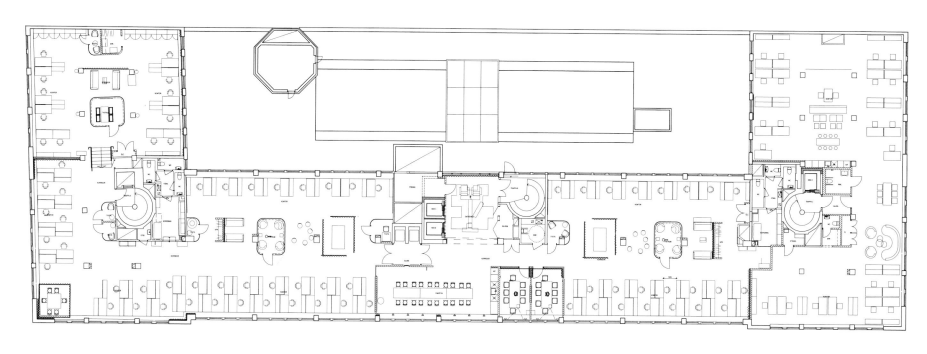

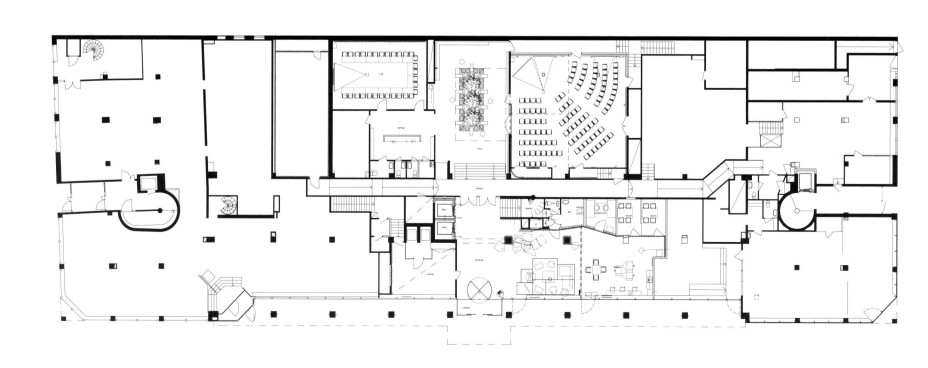

HSB是瑞典最大的住房合作社并由其成员所拥有。其位于斯德哥尔摩的办公室为了具备开放性和访问性，刚刚进行了一场彻底的改造。所有的这些改动都是为了申请环保级别即可持续发展评级体系银奖认证。pS 设计为本项目进行了内部设计和空间规划。为了通过声音和光线营造一种符合人体工程学的现代化办公空间，pS 付出了很多努力。设计结合现代科技，将分格式办公室变成开放式工作区域。

"社交互动"和"活力"成为关键词，设计主题为"欢迎回家"。接待区也叫做"生活商店"，与内部庭院相结合，让工作人员和来访客人可以随意地聊天。内部设计舒适而色彩缤纷，与原本 40 年代木造镶嵌的墙壁和电梯形成鲜明对比。

办公室空间的设计灵感来自于城市街区设计。每一部分都包括一些书桌，在中心区有一些给人们提供会面的"广场"和汇合点，并提供客厅家具、公用书桌和公用电话。

大概有 420 人在这栋大楼工作。顶层可以俯瞰城市的醉人美景，并提供数个室外会议室。休息室和一个叫做"智囊团"的创新空间是这个办公空间设计的画龙点睛之笔。

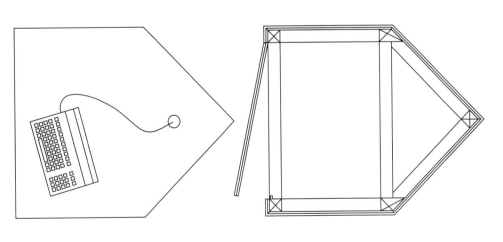
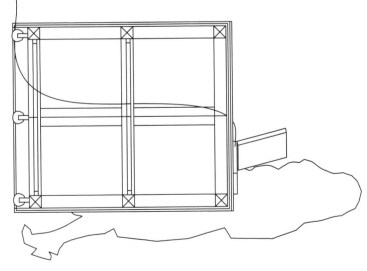

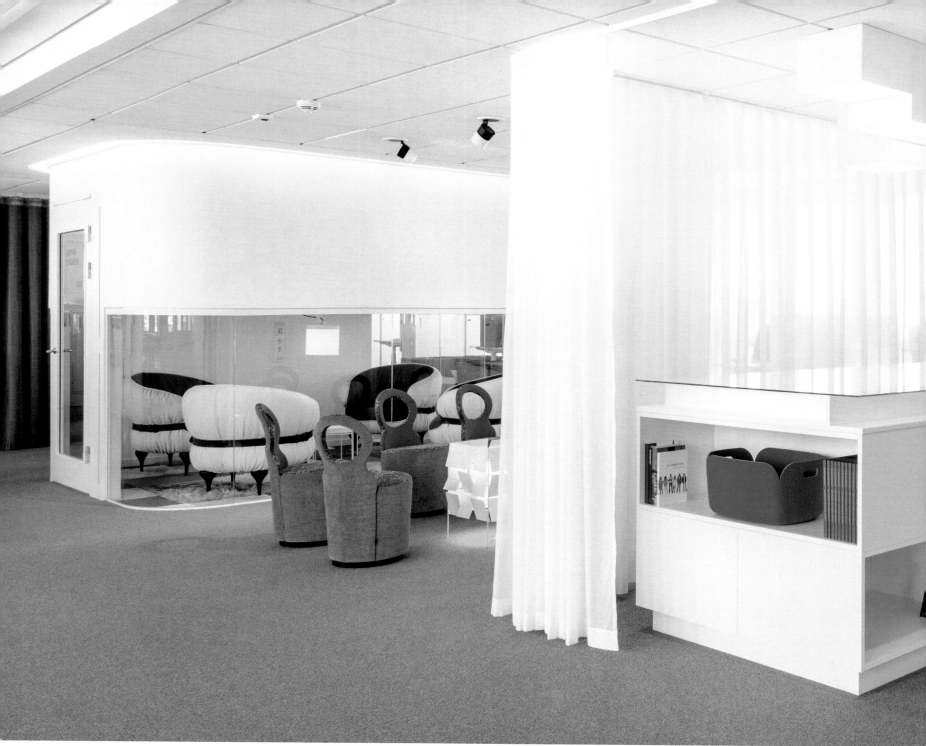

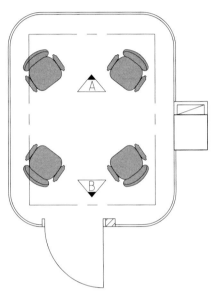

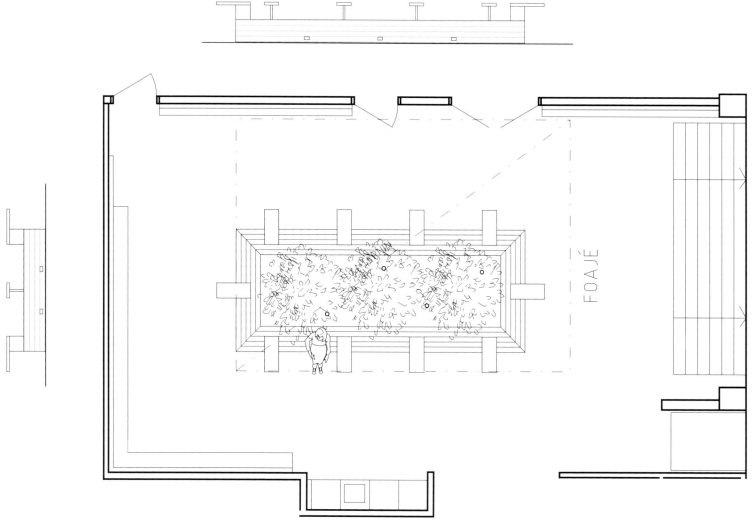

INNOCEAN HEADQUARTERS EUROPE

伊诺盛欧洲总部

Designer_ Ippolito Fleitz Group – Identity Architects
Design Team_ Andrew Bardzik, Anke Stern, David Schwarz, Frank Peisert, Sebastian Tiedemann, Yuliya Lytyuk, Gunter Fleitz, Peter Ippolito, Daniela Schröder, Tim Lessmann / Client_ Innocean Worldwide Europe GmbH
Location_ Frankfurt am Main Germany / Area_ 2,800 m² / Photographer_ Robert Hoernig

THE INTERNATIONALLY OPERATING ADVERTISING AGENCY INNOCEAN WITH HEADQUARTERS IN KOREA HAS MOVED INTO NEW EUROPEAN HEADQUARTERS IN FRANKFURT AM MAIN.

COLORS MATCH
色彩搭配

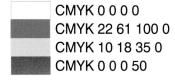

CMYK 0 0 0 0
CMYK 22 61 100 0
CMYK 10 18 35 0
CMYK 0 0 0 50

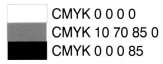

CMYK 0 0 0 0
CMYK 10 70 85 0
CMYK 0 0 0 85

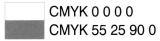

CMYK 0 0 0 0
CMYK 55 25 90 0

A flexible and modern work world was created for the young, design-conscious company, which fits the different work zones within the agency.

Dynamism and movement are key features of the design, which assimilates employees and visitors the second they enter the spacious reception hall. These design elements guide you through the open work zones and the specially created employee library right up to the in-house gym on the fifth floor, which offers an amazing view over Frankfurt. Polygonal spatial elements and a wide range of materials represent the high design standards of the agency itself. Open and transparent work areas, paired with semi-public and completely discreet conference zones promote a creative and communicative working atmosphere.

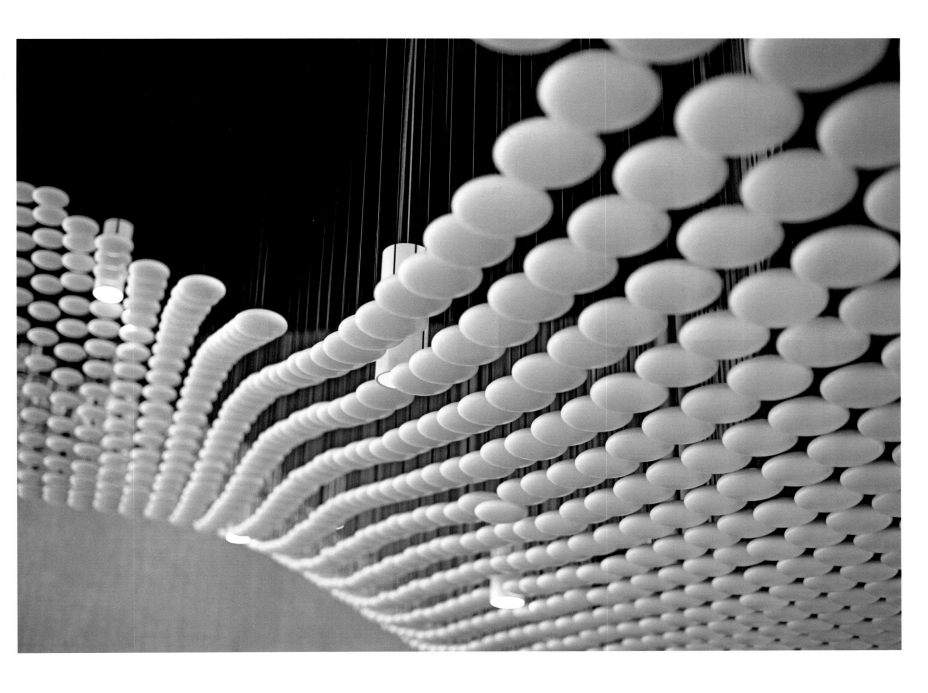

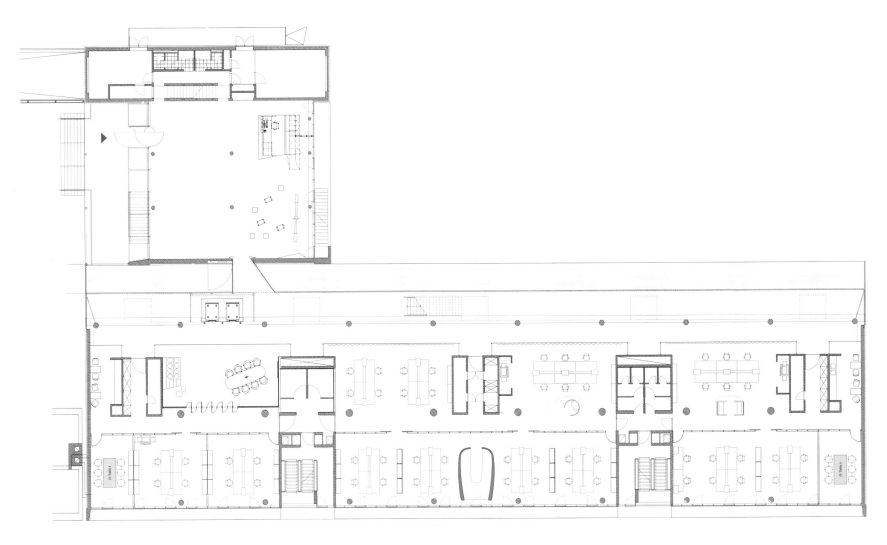

国际性广告运营公司伊诺盛在韩国的总部已经迁移至位于德国经济中心法兰克福的欧洲新总部。由于此公司年轻而具有独特的理念,因此新总部的设计采取灵活而现代的方式,以适应公司内部不同工作区的需求。

"活力"和"移动"是这项设计的主要特点,让员工和访客进入宽敞的接待大厅的一刹那就被深深吸引。这些设计元素引导你进入工作区域,以及五楼位于室内健身房旁边的图书室,在这里能够欣赏到法兰克福优美的景色。多边形空间元素和多样的材料运用代表了公司本身高端的设计标准。开放透明的工作区,与半公共区和完全封闭的会议区域搭配共同营造了一个创新和利于沟通的工作环境。

YANDEX SAINT PETERSBURG OFFICE – 4

圣彼得堡 Yandex 办公室 –4

*Designer*_Arseniy Borysenko, Peter Zaytsev / *Design Company*_za bor architects / *Decorator*_Nadezhda Rozhanskaya
*Project Management*_Yandex / *Client*_Yandex / *Furniture*_Herman Miller, Fritz Hansen, Walter Knoll, Pedrali
*Flooring*_Interface / *Location*_Saint Petersburg, Russia / *Area*_3,310 m² / *Photographer*_Peter Zaytsev

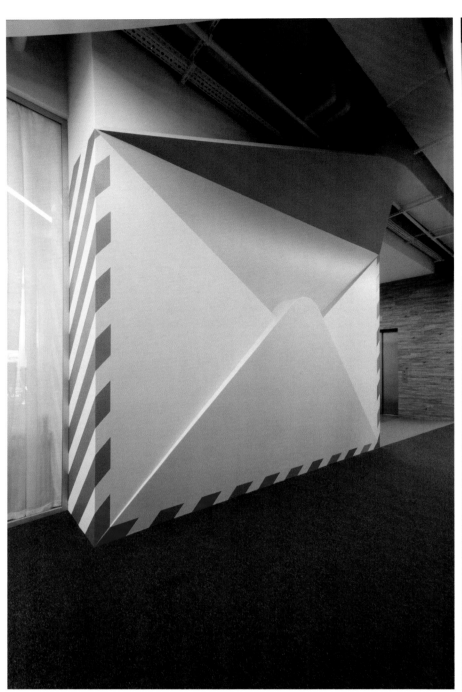

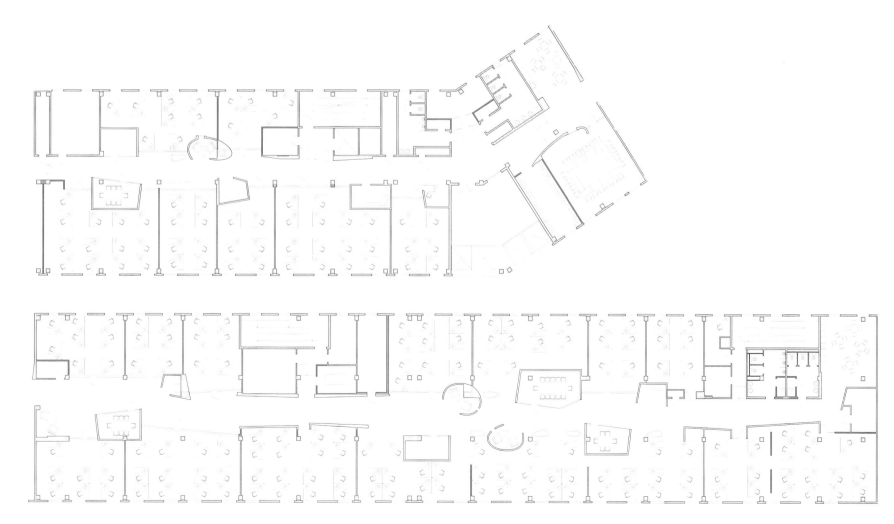

圣彼得堡Yandex办公室非常特别,许多原创的视觉方案使其成为全公司最明亮也最独特的办公室。随着最近几年公司的不断发展,需要更多的空间来设立办公室,因此2008年在圣彼得堡新建了1800平方米的办公区,到2014年办公室面积已经扩大到三层,大约10000平方米。最后一部分办公区,第四个办公室占据了Benua商务中心的三楼。

由于建筑师既熟悉商务中心区的空间也熟悉客户和建筑商的需求,因此工作进展得非常顺利。

客户想看到一个有趣的办公室,希望它在视觉上可以很好地结合位于五层非常明亮的部分和四层上安静的办公环境。

办公室设计充分利用了通过多年努力而研究出的最好的设计风格方案,并且在此期间搜索引擎设计不断发生变化,与客户公司象征符号相关的新的设计方案已经出现。现在Yandex已经有了天翻地覆的变化,许多平面图标都已经变成3D图标,通过在墙上切割而呈现"轮廓"的表现形式已经出现,这种切割法创造出了上百万搜索引擎用户所熟知的可辨识的符号轮廓。因此,根据建筑师的设计方案,办公客户和员工会全心致力于Yandex服务中,通过此服务,他们可以在二维平面上正常地互动。在最后完工阶段也投入使用了生态环保型的材料,例如,工业地毯,石膏板,其中一面墙壁还用稳定的苔藓覆盖起来作为装饰。

由于商业中心独特的规划设计,办公室将沿着近200m的走廊延伸,有足够的用于协商、谈判的房间和非正式沟通交流区域。由于这里人们24小时工作,为了方便员工,办公室还设立了淋浴间、餐厅、咖啡厅和休闲区,为员工创造了轻松、愉快的氛围。

ALFA BANK OFFICE

阿尔法银行办公室

Designer_IND Architects Studio / **Construction**_RD Construction / **Location**_Moscow, Russia
Area_2,630 m² / **Photographer**_Andrey Jitkov, Alexey Zarodov

ONE OF THE UNIQUE FEATURES OF THE OFFICE DESIGNED BY ARCHITECTS OF IND ARCHITECTS FOR ONE OF ALFA BANK'S BRANCHES IS, FIRST OF ALL, ITS CREATIVITY WHICH IS INCREDIBLE FOR THE FINANCIAL SECTOR.

COLORS MATCH
色彩搭配

CMYK 8 6 7 0
CMYK 35 44 79 10
CMYK 25 96 100 21
CMYK 70 63 70 75

CMYK 8 6 7 0
CMYK 1 26 82 0
CMYK 64 62 65 56
CMYK 84 70 22 5

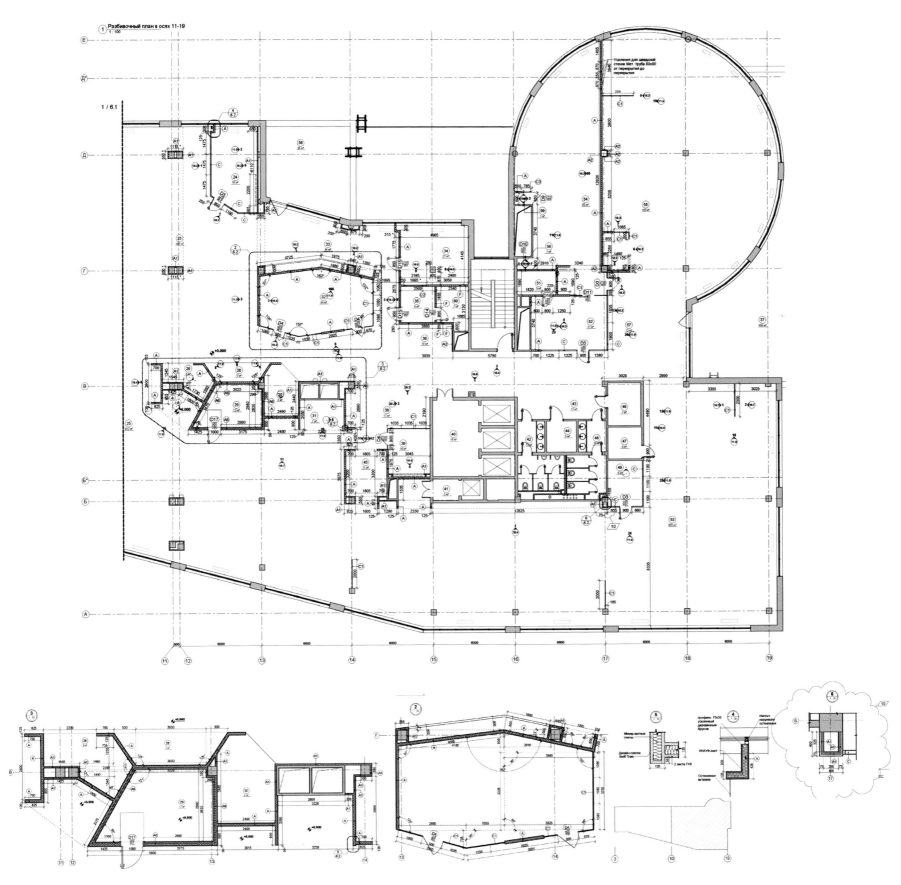

Functional, striking, and innovative, it will be a place to work in for young and vigorous employees of Alfa Laboratory — a special unit of Alfa Bank engaged in electronic business.

The idea of the interior is based on superheroes and street art — the components which Laboratory employees can relate themselves to. The finishing materials used are as follows: textured concrete, wood, perforated metal, various kinds of glass — clear, dim, and patterned. A bright carpet tile facilitates the navigation — meeting zones are colored, while various circles in an open space zone help employees to find required groups and departments.

Walls of the office combine several functions: a decorative function — bright wall murals featuring superheroes and comic books; a practical function — a special surface where you can write with felt-tip pens; a bulletin board material has been pasted on some walls, where employees can fix their materials; and an informative and motivational function — a wall with quotes of great people.

A distinguishing feature of the Laboratory is that employees can not only work on their work places, but take an active part in brainstorms and meetings as well. This feature has identified the laying out of the office — there are many meeting zones and buzz session zones (coffee points); the game zone to have a rest with a ping-pong table, various board games, and carpet-covered walls to play darts; and two outdoor porches with gleamy furniture here.

Various lighting solutions have been implemented in the office — linear light in the open space zone, LED backlighting in a hallway, and soffits in the game and the presentation zones. The classy, dynamic, and functional design with striking elements and interesting details — this is not merely an office, but a really comfortable place to create unusual and contemporary solutions, too.

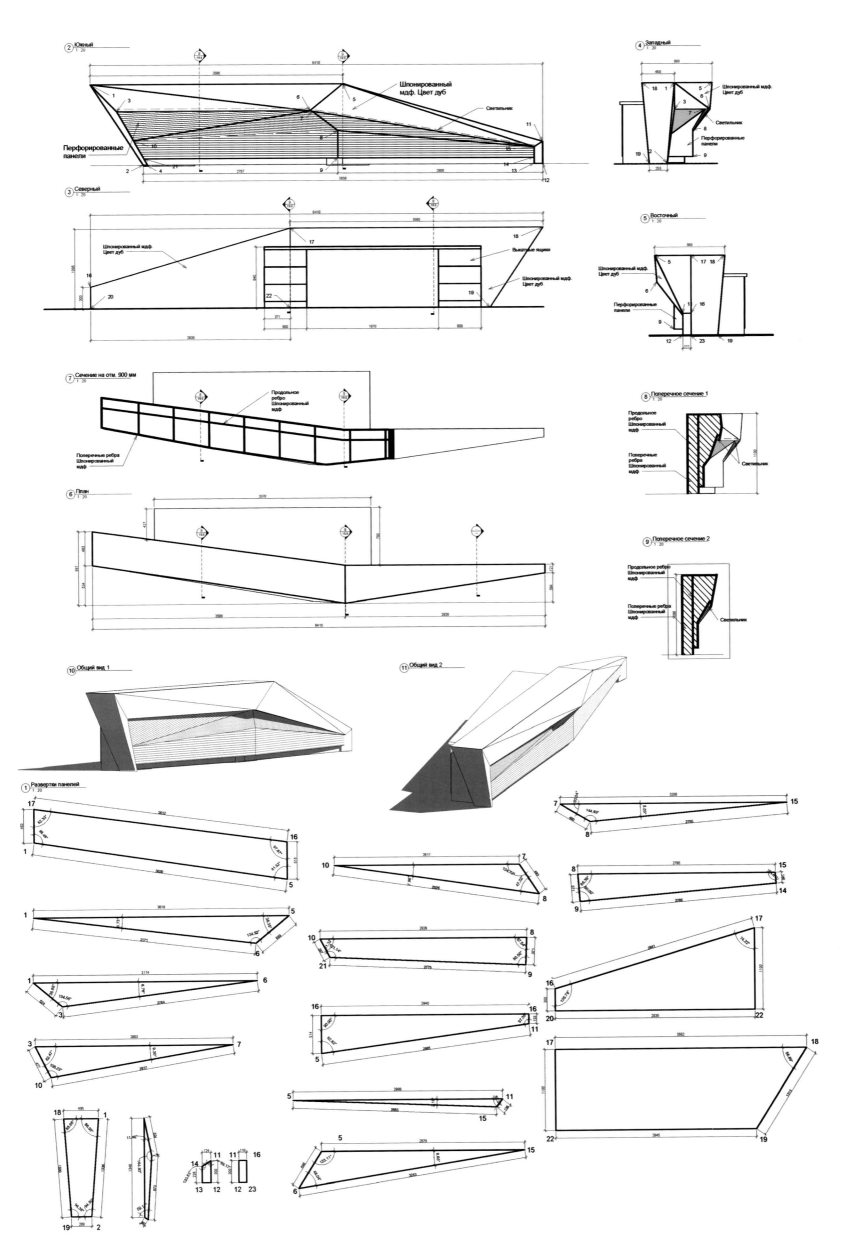

FLOOR EXPLICATION:
1. OPENSPACE
2. COFEE POINT
3. WARDBRODE
4. PRINT ZONE
5. STORAGE
6. MEETING ROOM
7. KITCHEN
8. RELAX ROOM
9. RECEPTION
10. CABINET
11. VIP RECEPTION
12. SERVER
13. WC

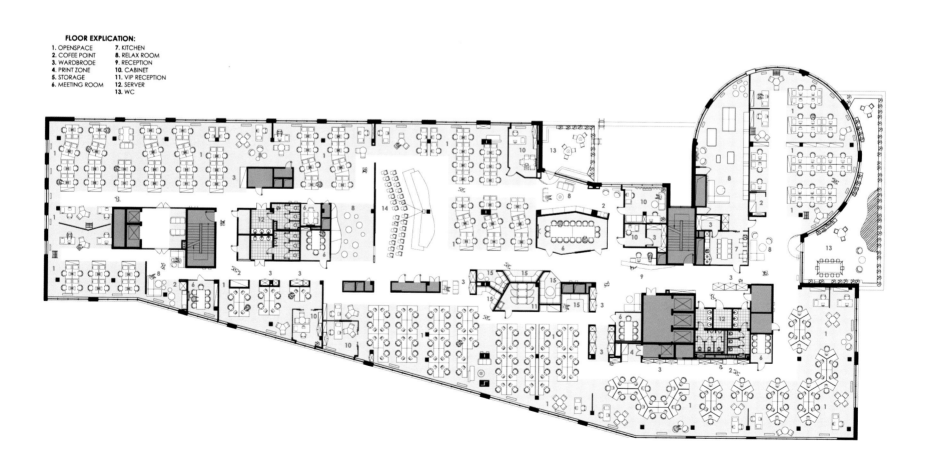

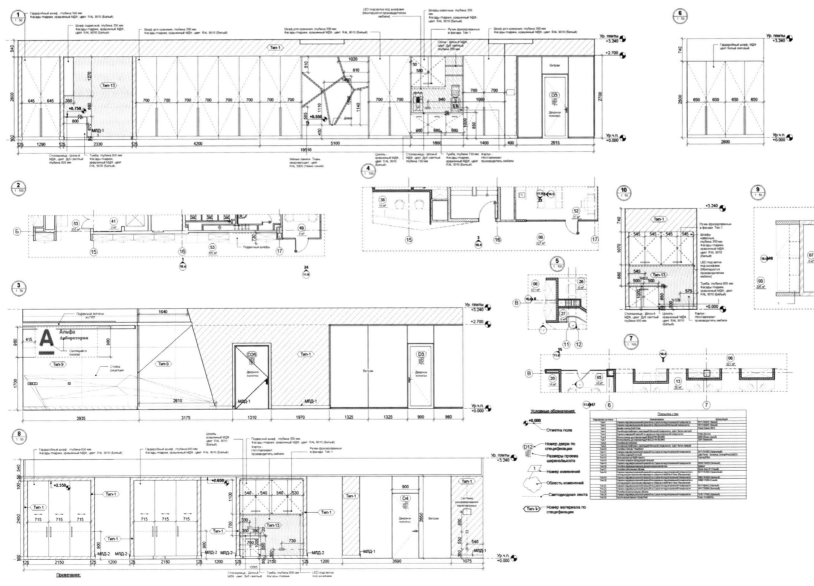

IND 设计事务所为阿尔法银行支行设计的办公室的独特之处首先在于它的创造性，这对一个金融部门而言是难以置信的。办公室功能齐全、醒目并独具一格，它将成为阿尔法研究所朝气蓬勃的年轻员工工作的好地方——阿尔法研究所是阿尔法银行从事电子商务业务的特殊部门。

内部设计的创意灵感源于超级英雄和街头艺术——研究所员工将自己和这些艺术要素联系起来。使用的涂饰原料包括：纹理混凝土、木材、穿孔金属以及各种玻璃——透明的、不透明的和有图案的。明亮的地毯纹理便于将人们指引到不同区域——会议区域是彩色的，开放空间区域的各种圆圈帮助员工找到相应的群组和部门。

办公室墙壁集合了几个功能：装饰功能——明亮的壁画以超级英雄和漫画形象为特色；实用功能——特殊的表面可任其使用签字笔在上面书写；墙面上贴上了像公告板一样的材料，员工可将资料固定在上面；教育和激励功能——墙面上写有伟人的名言。

该研究所的一个显著特点是，员工不仅可以在自己的工作场所工作，而且可以积极参与到集体讨论和会议中去。这个特点决定了办公室的格局——有许多会议区、小组讨论区（设有咖啡点）、游戏区（员工可以在此处玩乒乓球、各类桌上游戏，在地毯包覆的墙上玩飞镖），还有两个配有闪光家具的露天门廊。

该办公空间采用了不同的照明方案——开放空间使用线性光线，走廊、游戏区的底部和展示区使用 LED 背光源。拥有漂亮、活泼、功能性的设计，配有醒目的元素和有趣的细节设计——这不仅仅是一个办公室，还是一个真正舒适的地方，能创造出奇特而现代的新思维。

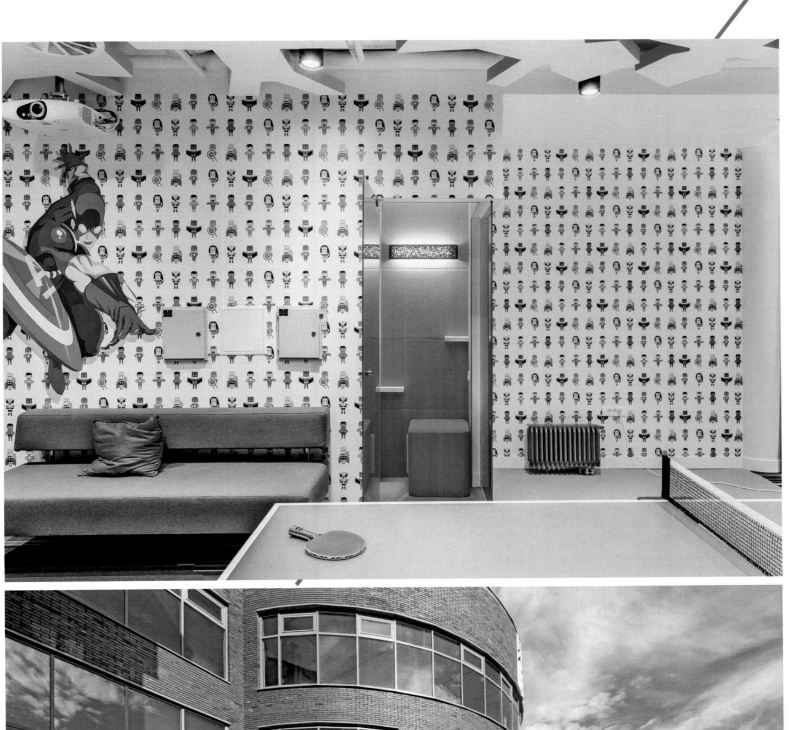

DROGA5

Droga5 办公空间

*Designer*_Rogers Partners Architects + Urban Designers
*Location*_New York, U.S.A / *Area*_9,290 m²
*Photographer*_Albert Vecerka/ESTO

ROGERS PARTNERS' DESIGN IS A COMBINATION OF OPEN OFFICES, PRIVATE OFFICES, WORKSTATIONS, AND MEETING ROOMS THAT ADDRESSES THE DIVERSE NEEDS OF FOUR DEPARTMENTS AND SUPPORT STAFF.

COLORS MATCH
色彩搭配

CMYK 0 0 0 0
CMYK 15 15 20 0
CMYK 0 0 0 100

CMYK 0 0 0 0
CMYK 80 45 30 0
CMYK 0 0 0 100

CMYK 0 0 0 0
CMYK 0 0 0 55
CMYK 0 0 0 100

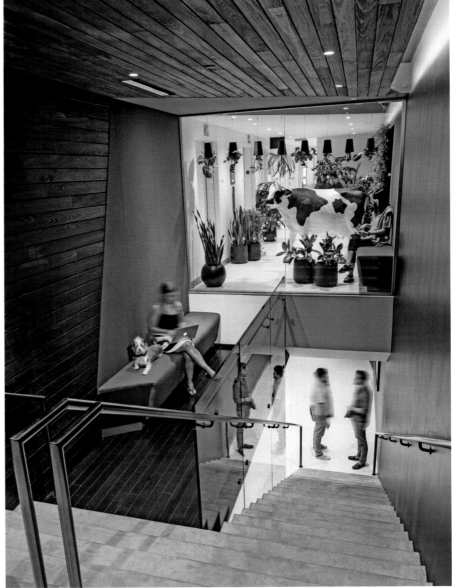

When Droga5, the dynamic, international advertising agency, decided to move their headquarters – and over 250 employees – from NoHo to Wall Street in New York's Financial District, they wanted a space that embraced their upstart beginnings while allowing room to mature as a company.

Common areas include a kitchen and dining area, reception, casual meeting spaces and a place for firm-wide meetings. In addition, a "work retreat" offers a private setting for client presentations and internal meetings.

To reflect the firm's collaborative culture, the traditional hierarchical design where one's place in the building is determined by status in the company was replaced with a "constellation" model with "attractors". Interaction is encouraged through chance encounters and impromptu meetings.

The vertical connections within the space provide opportunities for collaboration and places to gather. At the core, a large stair invites people to common areas below and serves as an amphitheater for agency meetings and social events.

The primary path connecting the floors is identified by the use of charred wood siding, providing way finding and a rich material quality to these areas. Simple authentic materials offer a sober and modern contrast with the unobstructed bright natural light flooding in from the East River views. Offices on the north and south perimeter are made of a translucent polygal, allowing light to be shared with the common spaces.

Upper floors feature the firm's "work retreat" away from day-to-day distractions; a corporate boardroom for meeting with potential clients; and an entertainment space for dinners and parties.

Keeping within the spirit of Droga5's creative culture, the space is designed to reject many of the traditional corporate office models, providing shared spaces that allow for interaction and flexibility and a capacity to expand and evolve with the firm into the future.

Rogers Partner 的设计将开放式办公室、私人办公室、工作站和会议室结合在一起，以满足四个部门和员工的不同需求。

当充满活力的国际广告公司 Droga5 决定将他们的总部——包括 250 多名员工——从曼哈顿以北搬迁到位于纽约金融区的华尔街，他们想寻找一个可以体现公司的高起点并且有足够的空间让公司发展得更加成熟的地方。

设计的公共区包括厨房、用餐区、接待、临时会议区和一个可供全公司开会的区域。此外，一处"休闲区"可提供专门空间以用于客户展示及举行内部会议。

为了体现企业的合作文化，以前公司的座位是由传统的上下级制度所决定的，现在改为非常受欢迎的"星座"模式。偶然邂逅和即时会议促进了员工之间的互动。

空间内垂直方向的衔接为相互协作和聚会提供了机遇和地点。在核心办公区，一个大型的楼梯可使人们进入下面的公共区，也可作为一个举办会议和社交活动的圆形会场。

连接楼层的主要路径使用了炭化木材壁板，起到了引路的作用，材质优良。简单而优质的材料形成沉稳的现代色调，与从东河方向射进来的强烈而明亮的自然光线形成鲜明的对比。南北向办公室边界由半透明魄丽佳阳光板构成，以便让公共区也充满阳光。

高层以"休闲区"为特色，远离日常干扰；并设有会议室，专门用于接待潜在客户，以及一个可供用餐和聚会的娱乐场地。

为了继续保持 Droga5 的创新文化精神，整个空间的设计摒弃了许多传统公司的办公模式，提供了可以互动，具有灵活性，未来拥有扩大和发展能力的共享空间。

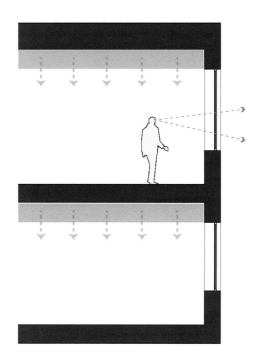 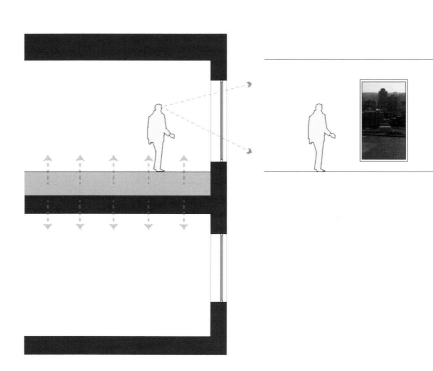

EXISTING/TYPICAL CONDITION   PROPOSED CONDITION

PARTY/PITCH/RETREAT

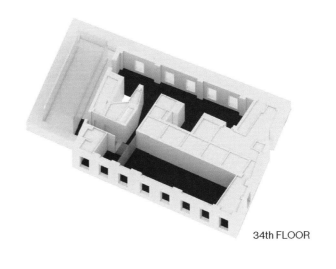

34th FLOOR

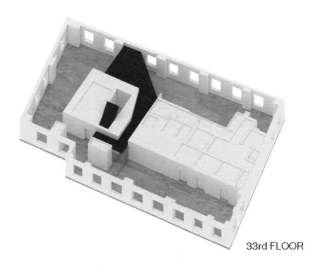

33rd FLOOR

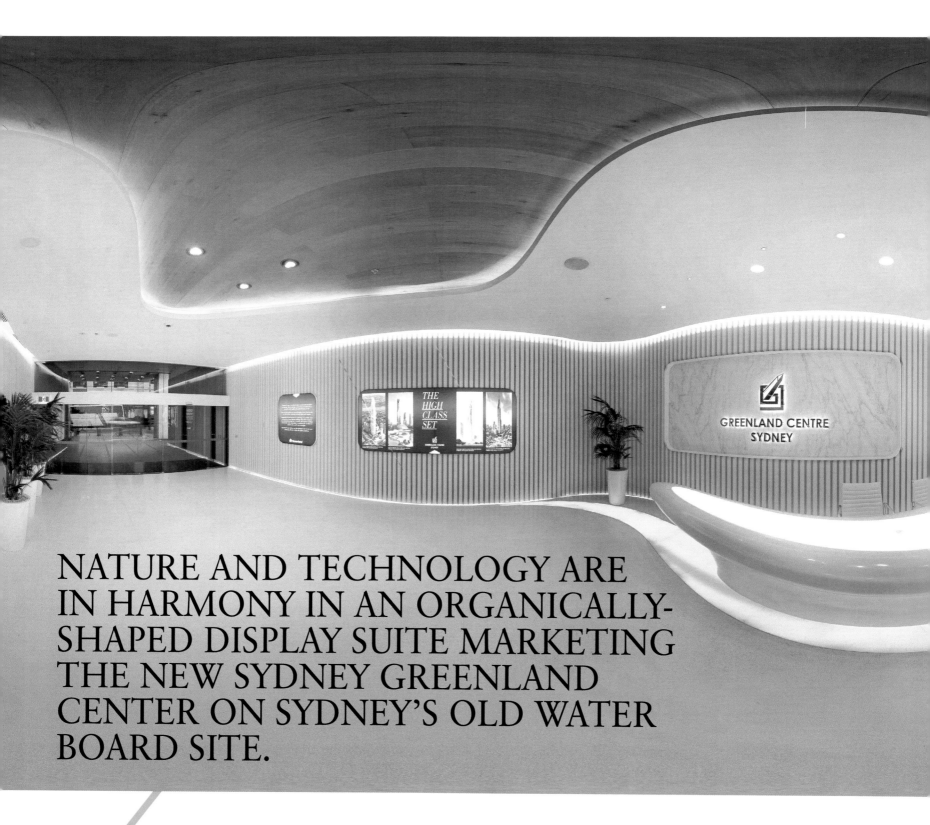

NATURE AND TECHNOLOGY ARE IN HARMONY IN AN ORGANICALLY-SHAPED DISPLAY SUITE MARKETING THE NEW SYDNEY GREENLAND CENTER ON SYDNEY'S OLD WATER BOARD SITE.

GREENLAND CENTER

绿地中心

*Designer*_PTW, LAVA

*Design Team*_PTW - Simon Parsons, Tony Rossi, Alex Lin, Joanna Varettas, George Freedman, Mark Giles; LAVA - Chris Bosse, Tobias Wallisser, Alexander Rieck with Jarrod Lamshed, Angelo Ungarelli, Giulia Conti

*Client*_Greenland Group / *Partner*_PTW / *Location*_Sydney, Australia / *Area*_450 m²

*Photographer*_Brett Boardman, Peter Murphy

The lobby of the 1960s Water Board building has been transformed with freeform furniture and curvy walls and ceilings to take the visitor through a journey of the new building.

LAVA partnered with PTW to create the display suite for the new mixed-use tower by the Greenland Group, which merges natural materials with high tech fabrication technologies.

The fluid space features white terrazzo floors, illuminated timber desks, and walls lined with white leather and timber battens. Continuous lighting ribbons create a luminous and airy environment.

The latest technologies include GRP (glass-reinforced plastic) — a lightweight, strong material that can be formed into fluid shapes. Parametric modelling and rapid prototyping means the design went straight from a 3D computer model to the fabrication workshop where the reception and display desks were CNC cut and coated.

LAVA believes that people in the 21st century are looking for spaces that link them to nature, and the forms found in nature — waves, canyons, clouds — create beautiful, efficient and connective spaces.

The Greenland Center will house apartments, retail and commercial. It incorporates two Water Board buildings: the adaptive reuse of the eight-storey heritage-listed 1930s Pitt Street building and a new tower on top of the 1965 Bathurst Street building.

COLORS MATCH
色彩搭配

```
CMYK 0 0 0 0
CMYK 10 10 12 0
CMYK 13 18 28 0
```

```
CMYK 0 0 0 0
CMYK 10 10 12 0
CMYK 68 43 10 0
```

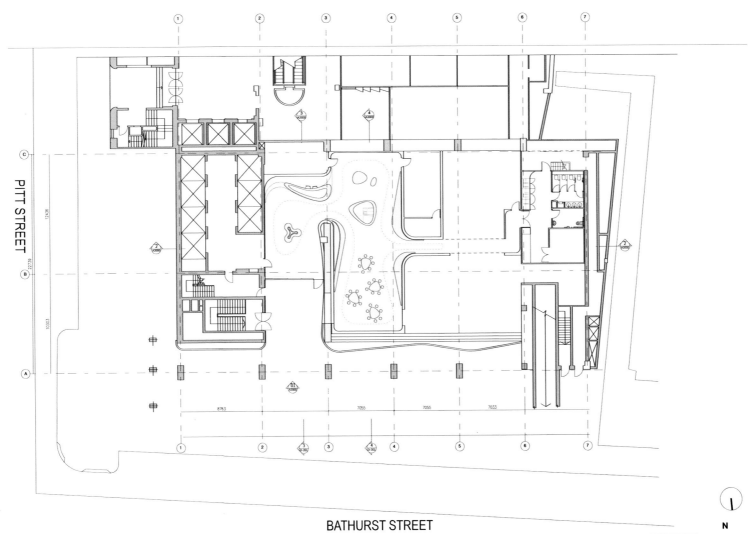

Ground floor plan

PITT STREET

BATHURST STREET

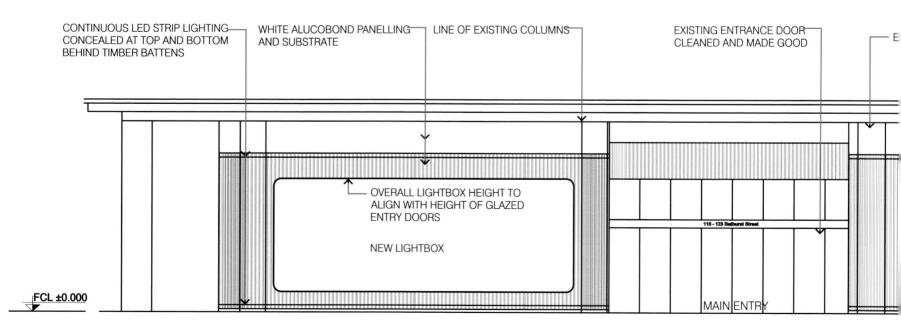

FRONT ELEVATION

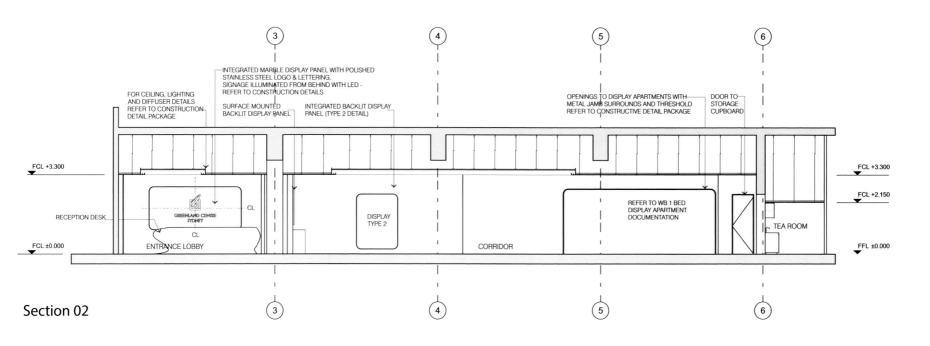

Section 02

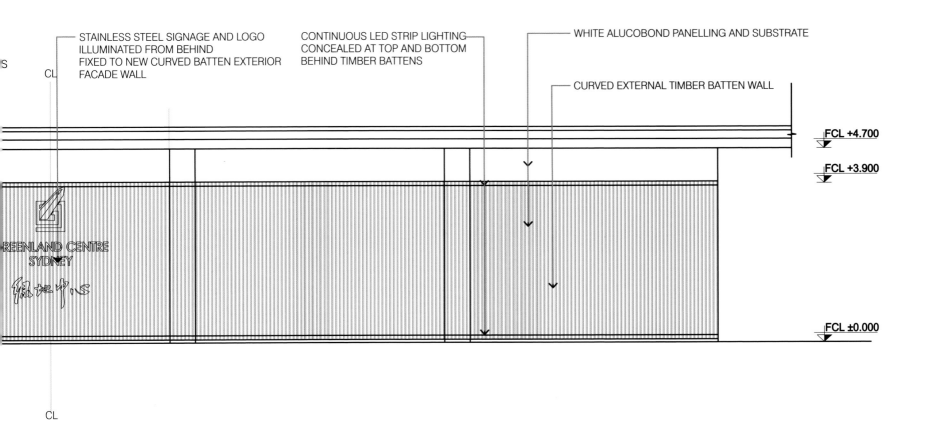

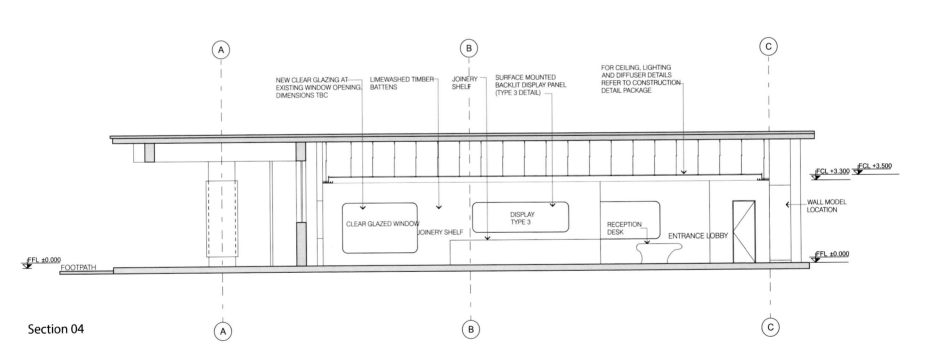

Section 04

　　在悉尼的 Water Board 旧址上，自然和科技有机而和谐地体现在新悉尼绿地中心的样板空间中。
　　这个建于 20 世纪 60 年代的 Water Board 大楼的大厅利用天然材料并通过家具自由组合、弧形墙壁和墙顶天花的顺势弯曲带给参观者一段全新的大楼之旅。
　　LAVA 和 PTW 利用自然材料通过高科技制造技术，合力为绿地集团投资的全新综合公寓塔楼打造样板房。
　　流线型的线条轻巧又精致，白色的水磨石地面、灯光照射的木质桌椅、白色的皮革和木质的压条装饰的墙壁使其十分别致，连续的灯带营造了一种透亮而梦幻的氛围。
　　大楼的最新科技包括 GRP（玻璃钢），即一种轻型而坚固材料，可以塑造出流线型空间。参数模型和快速成型技术的运用意味着设计从 3D 计算机模型直接到达金属加工工作室，在那里用于接待和展览的椅子将进行数控切割和涂层。
　　LAVA 坚信，21 世纪的人们渴望接近自然和自然界中的海浪、峡谷、白云，建于共同构造出美丽、有效而和谐的空间。
　　绿地中心集公寓、零售和商业于一体。它整合了两幢 Water Board 大楼空间，分别是建于 20 世纪 30 年代的 8 层高的 Pitt 大街老楼和建于 1965 年的一幢 Bathurst 大街大楼顶端的新塔楼。

LEASEPLAN

Designer_Rosan Bosch Studio / Client_LeasePlan
Location_Brøndby, Denmark / Photographer_Rosan Bosch, Kim Wendt

LeasePlan 办公空间

LEASEPLAN IS LEADING AT THE DANISH CAR LEASING MARKET. NOW, THE COMPANY HAS RECEIVED A PHYSICAL DESIGN THAT SUPPORTS ITS MARKET SHARE.

Rosan Bosch Studio has set shape and color on LeasePlan's values and created a design that both puts the customer first and optimises the workflow for the employees.

In LeasePlan's new showroom, an open reception desk greets the customer and emphasises the message: Here the costomer is in focus - not the car. The colors of the company's logo characterise the interior and create a beautiful and exclusive framework for the showroom.

Rosan Bosch Studio's design for LeasePlan's new headquarters focuses on the meeting and communication with the customer as well as the values that characterise the company. This not only applies to customer facilities but also to the underlying administrative buildings.

With identity graphics, custom designed meeting facilities and inclusive public spaces, a bridge between showroom and office areas is created. The office areas convey the same participatory experience for the staff as a showroom does for the customers. In this way, the development and design project gathers the aims of LeasePlan to prioritise the best customer experience, while at the same time creating a better workplace for the employees.

The reform of the Microsoft headquarters in Madrid, located in La Finca. Flexible work forms were implanted to a population of 720 professionals. The new system frees up 3,000 of the 9,000 m² that had been occupied by the company for 10 years.

The Microsoft company has been implementing flexible forms of work by Workplace Advantage program for more than five years and, even better, the protocol of the company states that all headquarters reform must shift towards this model of unallocated space. For this it makes available to its employees staff management and technology tools that enable them to work wherever and whenever they want with clear rules for evaluating performance and fulfilment of the objectives.

The savings in space meant reducing the area dedicated to traditional workplaces permitting putting up a new client area. The ground floor would be designed to promote the Microsoft products through the experience transferred by technology and the characteristics of the space.

The challenge for 3g office was to create a space that would invite to return to it, and that the company headquarters remained the favourite meeting place. For this, around two thirds of the surface of the work zones is intended for collaborative spaces of all kinds. Flexible forms of work permit to work from, where and when the worker wants, also within the office.

COLORS MATCH
色彩搭配

CMYK 0 0 0 0
CMYK 20 20 23 0
CMYK 35 50 65 1

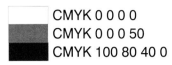
CMYK 0 0 0 0
CMYK 0 0 0 50
CMYK 100 80 40 0

CMYK 0 0 0 0
CMYK 0 15 100 0
CMYK 35 50 65 1

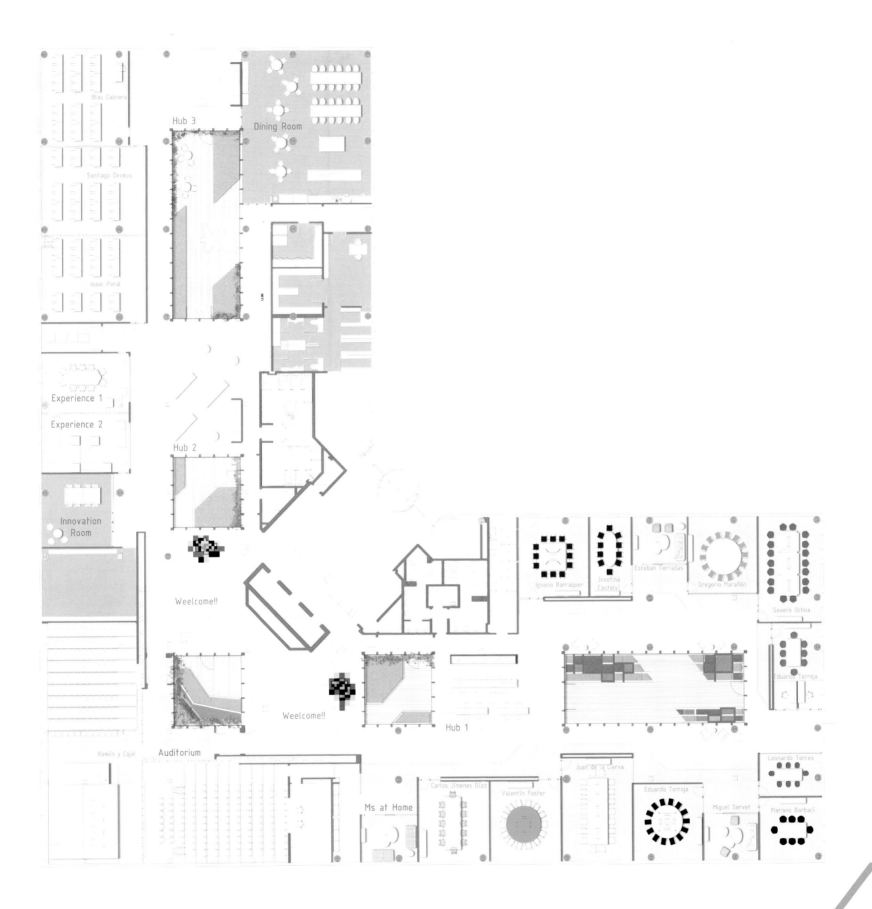

灵活的工作形式和空间利用的最大化一直是微软对其新总部的要求，新总部应有别于旧的空间，并且在不增加租用面积的前提下，创建一个新的客户区。

改建的微软总部位于马德里的 La Finca 酒店。灵活的工作形式深深影响着公司 720 名专业员工。10 年来公司规模达到占地 9000 平方米，新的体系为公司节省了 3000 平方米。

微软公司凭借工作场所优势实施灵活工作形式已有五年多了。公司的议定书中声明所有总部改建必须遵循空间不分配模式。这样有利于员工的管理和技术工具的使用，使员工在明确有关评估表现和目标完成情况的规章制度下不受时间地点限制地工作。

空间的节省意味着缩减传统工作场所的面积，这样就可以搭建一个新的客户区。第一层的设计旨在通过技术转让体验和空间特点来推销微软产品。

3g office 设计团队面临的挑战是要创造一个人人都想再来的空间，使公司总部依旧成为最受人们喜爱的会面场所。为了实现这一目标，工作区大约三分之二的空间被打造成各种各样的合作式空间模式。灵活的工作形式使得员工可以在办公室内或公司内的任何地方、任何时间都自如地工作。

NOTTING HILL

*Designer*_Yunakov Architects
*Design Team*_Yunakov Ivan, Kornienko Olga, Karpov Andrey
*Location*_Kyiv, Ukraine / *Area*_170 m² / *Photographer*_Oleg Stelmah (Electraua)

诺丁山

LOCATED IN THE LOWER PART OF THE KIEV CITY IN UKRAINE, THIS PLACE WAS CREATED AS A MULTI ZONE SPACE.

As a platform for creative events practically in every sphere: photo studio, intellectual entertainments, workshops, lectures, concerts, trendsetter's debates.

The space is divided into 4 zones. Reception area, kitchen, makeup zone, Kensington Hall and Chelsea Hall.

Each zone is made in its own style not breaking the overall style and concept.

Design and appearance of Notting Hill combine conceptual and tribute to the Kiev urban context, functionality and perfectionism. In the design of space were used noble materials of the highest quality, and some elements received new life thanks to a non-trivial design findings and recycling ideas.

All this – the result of a collaboration of architects, painters, sculptors, and designers.

COLORS MATCH
色彩搭配

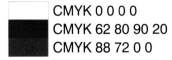
CMYK 0 0 0 0
CMYK 62 80 90 20
CMYK 88 72 0 0

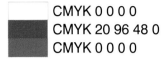
CMYK 0 0 0 0
CMYK 20 96 48 0
CMYK 0 0 0 0

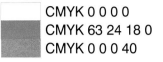
CMYK 0 0 0 0
CMYK 63 24 18 0
CMYK 0 0 0 40

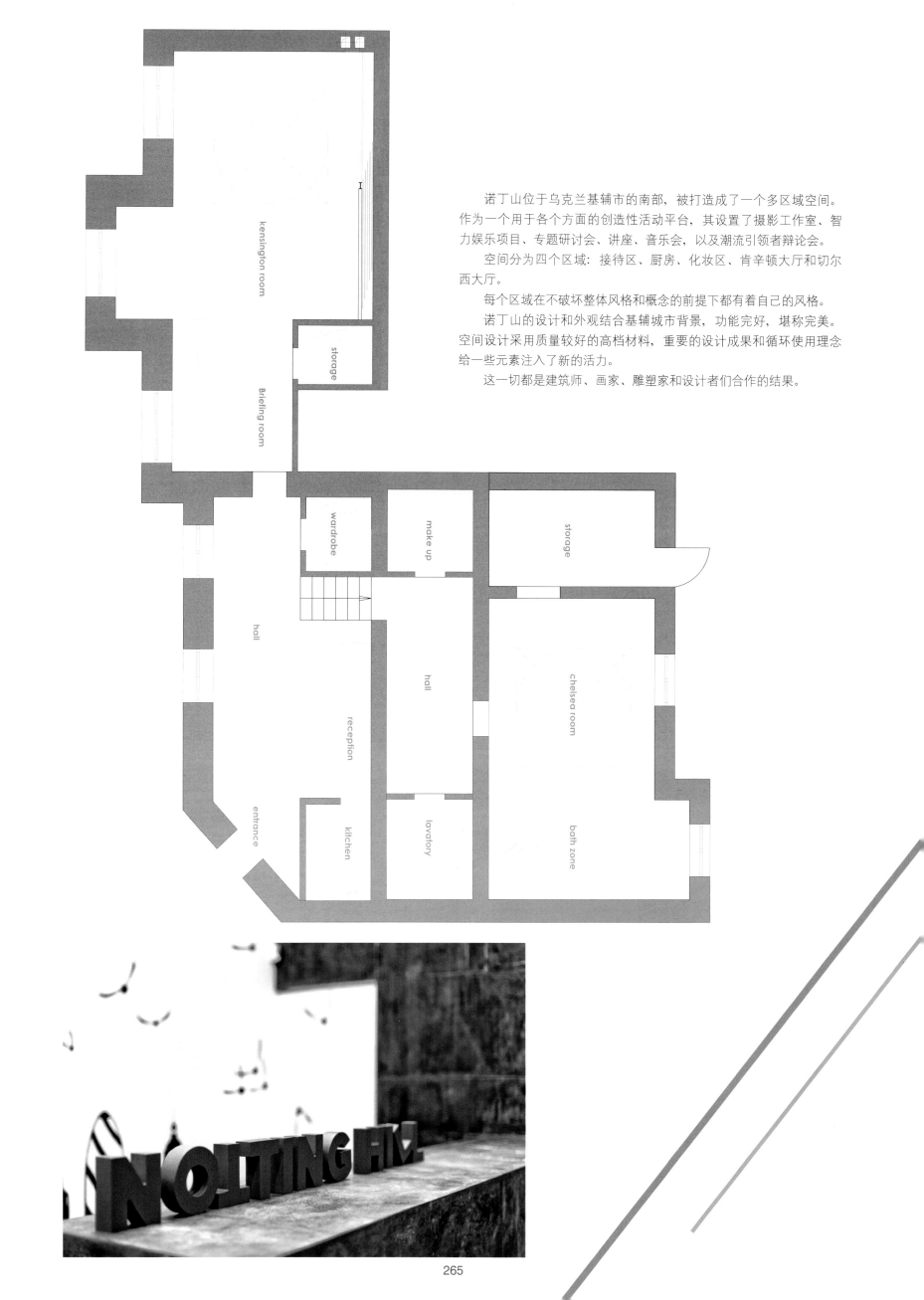

诺丁山位于乌克兰基辅市的南部，被打造成了一个多区域空间。作为一个用于各个方面的创造性活动平台，其设置了摄影工作室、智力娱乐项目、专题研讨会、讲座、音乐会，以及潮流引领者辩论会。

空间分为四个区域：接待区、厨房、化妆区、肯辛顿大厅和切尔西大厅。

每个区域在不破坏整体风格和概念的前提下都有着自己的风格。

诺丁山的设计和外观结合基辅城市背景，功能完好，堪称完美。空间设计采用质量较好的高档材料，重要的设计成果和循环使用理念给一些元素注入了新的活力。

这一切都是建筑师、画家、雕塑家和设计者们合作的结果。

ONEFOOTBALL

Onefootball 办公室

Designer_TKEZ architecture & design / Client_Onefootball
Location_Berlin, Germany / Area_1,486 m² / Photographer_Benjamin A. Monn

MUNICH BASED ARCHITECTS TKEZ ARCHITECTURE & DESIGN HAVE DESIGNED AND BUILT A NEW HEADQUARTERS FOR THE WORLD'S LEADING FOOTBALL COMMUNITY ONEFOOTBALL.

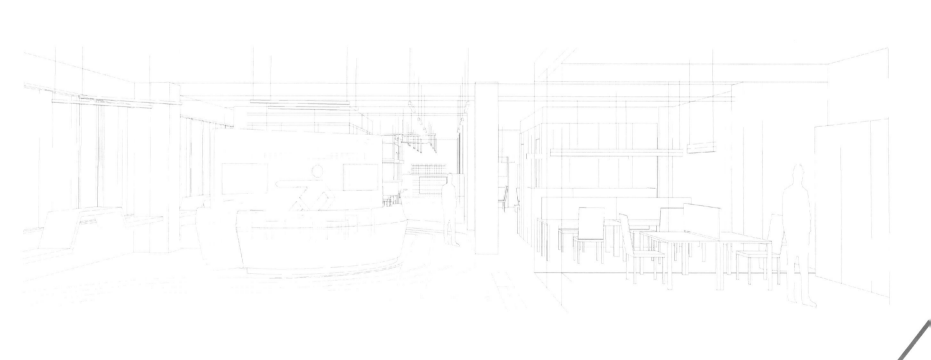

COLORS MATCH
色彩搭配

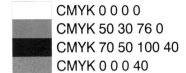

CMYK 0 0 0 0
CMYK 50 30 76 0
CMYK 70 50 100 40
CMYK 0 0 0 40

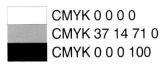

CMYK 0 0 0 0
CMYK 37 14 71 0
CMYK 0 0 0 100

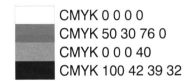

CMYK 0 0 0 0
CMYK 50 30 76 0
CMYK 0 0 0 40
CMYK 100 42 39 32

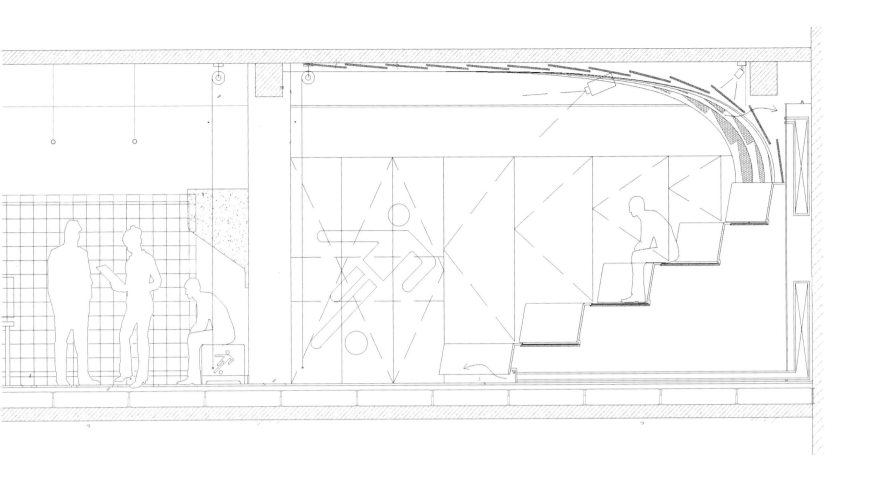

慕尼黑设计建造公司 TKEZ，为全球领先的足球公司 Onefootball 打造了一个新的总部办公室。

整个办公室开阔明亮，而且具有多重功能，代表了 Onefootball 公司朝气蓬勃、活力四射的精神。

动感的草绿色跑道呈现大气的地面铺装。

在这个宽敞的空间里，有带有透明落地玻璃的单人或者双人间、办公室以及会议室，使空间节奏感十足。为了给多样化的团队营造一种微妙的区划，这些场所都被精心地规划了。如果需要一个更加私人、安静的工作环境，还可以用半透明的窗帘来遮挡。

人造草皮地板参考了该公司的主题：足球来设计。

狭长的即用区域使员工可以离开自己的办公桌，到这里和同事开个临时会议或者面向充满阳光的绿色庭院工作。

"软工作"区是由纺织小屋构成的。办公室最后面的沙发区是大家见面闲聊的地方。墙上有特殊磁性的可反复书写的面板，可在集思广益的会议上派上用场。每个人的想法和想象都可以写在墙上，同事之间可以相互交流讨论。

草绿色跑道在球门处终止，在工作休息时员工可以进行点球大战。

新总部中间是 Onefootball 竞技场，这个多功能的区域是整个办公空间最主要的一部分，这里既是展示区、会议区和工作交互区，又可以用来发布足球赛事或者用于公司聚会。

在那里，电脑程序员或设计小组可以聚在一起开研讨会，在 4m×4m 的屏幕上展示最新的想法和工作情况。

在有着彩色背光屋顶和环绕音响设备的足球比赛竞技场上，到处都是公司员工，以及他们的家人和朋友，营造了一种独特而充满活力的氛围。

COMPULSIVE PRODUCTIONS

令人着迷的办公空间

*Designer*_Matt Gibson / **Design Company**_Matt Gibson Architecture + Design
*Project Team*_Matt Gibson, Phil Burns, Angela Hopkins, Carolin Arndt
*Builder*_Cubed Projects / **Location**_Melbourne, Australia / **Area**_216 m²
*Photographer*_Shannon McGrath

THE KEY STRATEGY OF THIS PROJECT WAS TO CREATE A MULTI-FACETTED WORKPLACE THAT ALLOWS CROSS-POLLINATION AND DIVERSITY OF EVENT.

Crafting flexible workplace arrangements that provide the owner with multiple avenues for rental revenue, allows the core business to gain increased exposure and opportunity for new work via association and collaboration with its tenants.

A key requirement was to introduce a state of the art film editing suite for the primary business, with potential to dry hire to third parties. Inspired by the geometry of the clients' tools of trade, the warehouse space was spanned by a series of open ended ply-clad acoustically lined pods — notionally emulating a camera barrel — allowing visual permeability and connectivity. Used for production reviews and color grading these spaces need to control natural light and transform to dark spaces.

Budget dictated that major change was out of the question. In a cost-effective strategy that brought the newer more sophisticated elements into the foreground, the remainder of the space is finished in various shades of black-paint or bronzed-mirror conveniently recessing and camouflaging existing elements.

Aligning with studies of sci-fi and film-noir genres, the concept delights in the opposition of darkness and light; and the tension of past, present and future. The futuristic pods — finely detailed, hi-tech and gleaming, enclosed and cocooning; play off the seemingly vast and infinite exposed nature of the existing warehouse — where surfaces are left rudimentary, old and industrial allowing a relaxedness and ability to personalize these left-over spaces.

At night, with ability for stage sets and events, the space transforms with low-key lighting and dark reflective surfacing; to be at once sumptuous, moody and mysterious. The corporate color red is interwoven only in detail — the company name does not appear once internally, instead red is subtlety woven through furniture, objet and neon artwork.

COLORS MATCH
色彩搭配

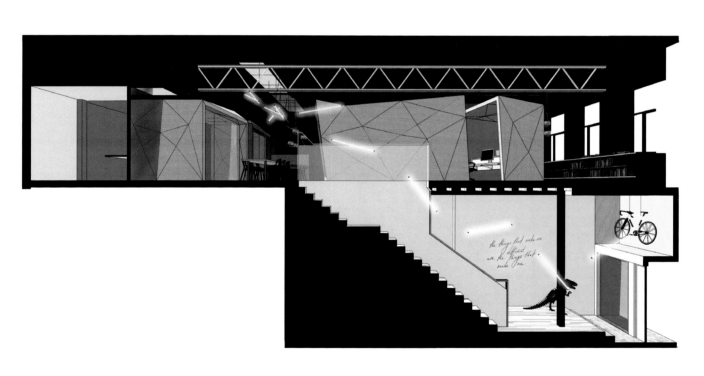
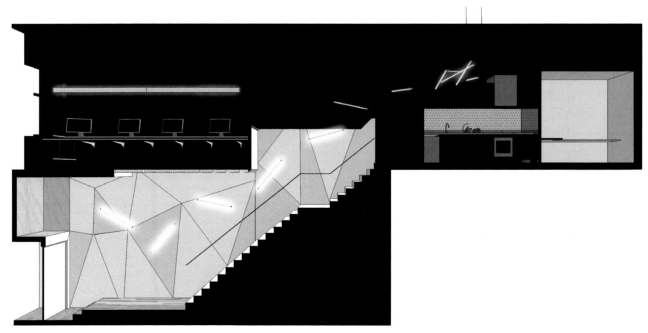

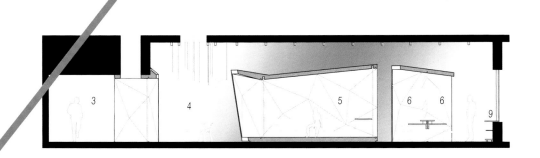

PROPOSED SECTION

3 SUB TENANT EDITING SUITE
4 BREAK OUT/MEETING SPACE
5 MAIN EDITING SUITE
6 PRODUCER/DIRECTORS DESK
9 LIBRARY/DISPLAY

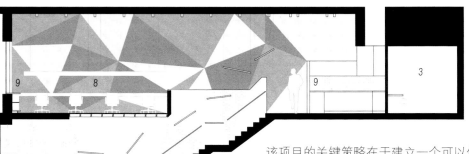

PROPOSED SECTION

1 ENTRY/GALLERY
3 SUB TENANT
7 HOT DESKS/WORK STATIONS
8 KITCHEN
9 LIBRARY/DISPLAY

该项目的关键策略在于建立一个可以促进彼此交流，并可以举办多种活动的多功能工作空间，设置弹性的工作场所，给所有者获得租金收入提供多种渠道，通过与租赁者的联合协作使核心业务获得更多的曝光率，并且能带来更多新的工作机会。

一个关键的设计要求是要为主要业务引进先进的电影剪辑套间，使之具备租给第三方的潜力。受顾客使用的贸易工具的几何形状的启发，仓库空间被设计成了一系列开放式并且带有声线界面的空间——理论上模拟镜头筒——达到视觉上的通透性和连通性。这些空间用于产品的评价和色彩分级，需要控制自然光，并将其转化为黑暗的空间。

预算决定了不允许做重大的改变。在成本效益策略指导下，同时保证在前景前添加更多更复杂的元素，空间其余部分的完成主要依赖于便捷的凹陷镶嵌以及掩盖现有元素的多种黑色漆或古铜色镜面。

与科幻和黑色电影流派研究相呼应，这个设计理念因光明与黑暗的对立、同时也由于过去、现在和未来的不断延伸而引人入胜。未来主义界面——极度精细、科技化、闪闪发光，封闭犹如蚕茧，暴露出现有的看似广阔庞大的工作室的本质——表面残缺、破旧并且工业化，因此需要放松的态度和较强的能力将这些破旧的空间个性化。

到了晚上，因为具有舞台布景和适应场合的能力，这些空间会随着低调的灯光和深色的反光表面而发生变化，马上变得华丽、变化多端和神秘莫测。公司使用的标志性红色穿插在细节里——公司名称没有使用红色，而是将红色精妙地融合在家具、装饰和霓虹灯工艺品中。

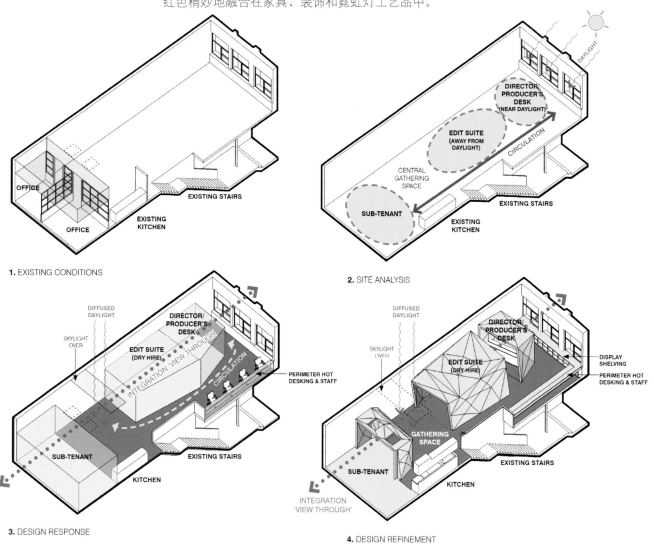

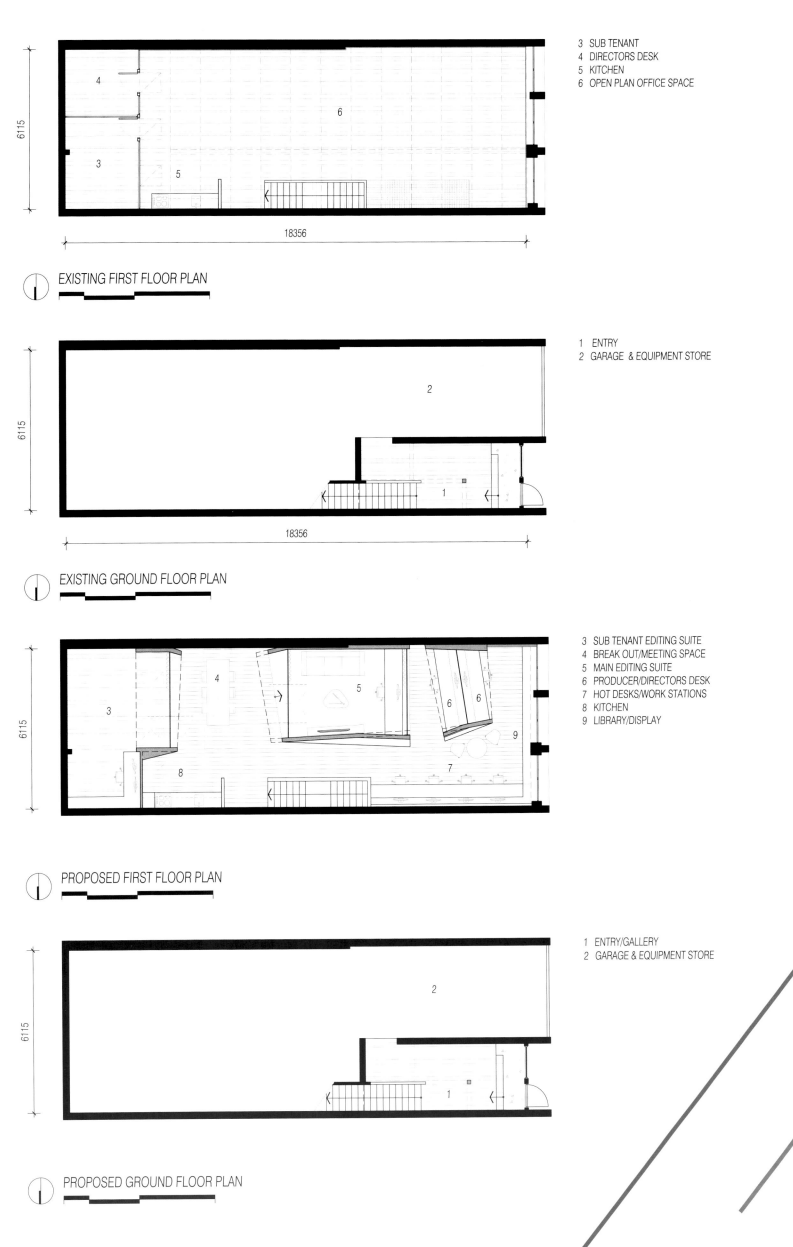

CUBIX OFFICE

Cubix 办公室

Designer_Kapil Aggarwal
Design Company_Spaces Architects@ka
Design Team_Kapil Aggarwal, Pawan Sharma
Site Supervisor_Arvind Pal Singh
Client_Cubix Homes / Location_New Delhi, India
Area_111.48 m² / Photographer_Bharat Aggarwal

THE OFFICE FOR REAL ESTATE CONSULTANT HAS BEEN CONCEPTUALIZED AS A MODERN WHITE OFFICE WITH FLUID FORMS.

The site being linear with 4m wide with a depth of 24m was a challenge to create individual cabins which were to placed one behind each linearly creating a corridor space connecting them, to avoid it the conference placed at the center of the space has been designed in elliptical oval form to have free flow also the cabin behind was designed with angled glass partition to connect corridor space with the interior space visually thus creating interesting movement spaces at the same time creating transition.

The front of the office being 3m with kitchen wall in fluid profile forms a backdrop for the reception. A building model designed has been placed vertically on the wall in front of reception adding character to the space. The reception table designed follows the fluid concept of the space with abstract backlit panels is in harmony with curved back wall which takes a peel form, the top curve panel extending towards workstation and other supported by angled column.

The conference room has multiple layered panels with glass slit, the form at the center acts as a transition dividing the office in public and semi-private spaces. The conference table has been designed by combining multiple curved panels fixed together with a glass top in elliptical shape. The corridor leading to the rear room has been designed with multiple project images display in black which with fluid form ceiling in abstract shape backlit panel reflecting the pattern on the floor also. The ceiling of the cabin behind the conference designed in curved profile with multiple grooves has an abstract shape ceiling hanged below. The fluid and abstract form extends to the furniture designed in the space. The flooring of MD room has black tile in contrast to the grey used outside to create transition, and the ceiling has been designed in fluid form with abstract shaped black painted panel's backlit.

COLORS MATCH
色彩搭配

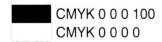
CMYK 0 0 0 100
CMYK 0 0 0 0

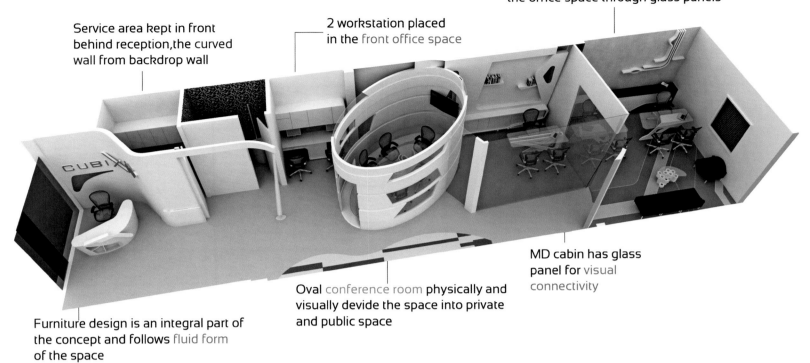

Service area kept in front behind reception, the curved wall from backdrop wall

2 workstation placed in the front office space

Rear MD cabin has dark flooring reflecting ceiling profile. The design intension is to visually connect the office space through glass panels

Furniture design is an integral part of the concept and follows fluid form of the space

Oval conference room physically and visually devide the space into private and public space

MD cabin has glass panel for visual connectivity

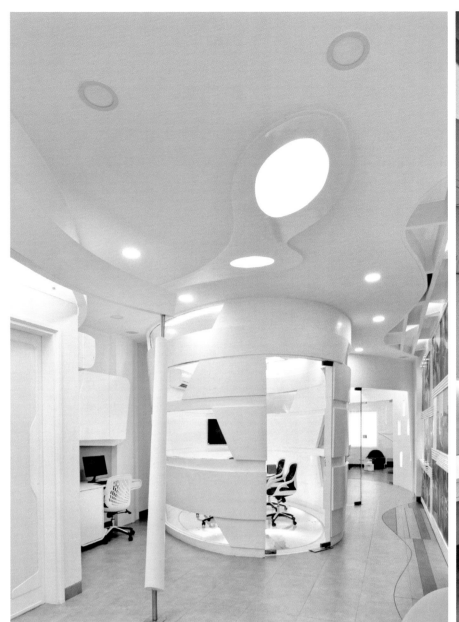

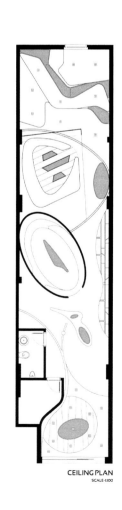

LEGEND
1. Entrance
2. Waiting Area
3. Reception
4. Pantry
5. Toilet
6. Workspace 1
7. Conference Area
8. Director's Room
9. M.D's Room

PLAN SCALE 1:100

CEILING PLAN SCALE 1:100

　　该房地产咨询办公室被定位为现代的白色流动型办公室。该狭长型场地4m宽、24m长。在两端各设置一个个人房间，形成一个走廊将两个房间连在一起的形式，这对于设计者来说是个极大的挑战。为了避开难点，将会议室设在了空间的中心位置，呈椭圆形，并可以自由地移动。同时，后面的卡间在设计中也加入了有角度的玻璃隔板，从而在视觉上衔接走廊与内部空间，这样就创造了有趣的动感空间，同时也营造了空间的过渡效果。

　　办公室前部尺寸为3m，厨房墙壁呈流线型轮廓，也用作接待区的背景。设计好的建筑模型垂直放置在接待中心前面的墙壁上，给空间增加了特色。在设计过程中，接待桌采用了流畅型空间概念，附有抽象的背光式展板，与曲线背壁和谐一致。顶部的曲形面板朝着工作站扩展，而其余部分被角柱支撑。

　　会议室使用带有玻璃缝隙的多层次墙板，中心部分作为一个过渡，将办公室划分为公共和半私人空间。会议桌的设计将许多弯板与椭圆形的玻璃表面固定在一起。通往后面房间的走廊被设计成展示项目黑白图片的展示区，走廊天花板的形状抽象而流畅，背光面板也可将图案投射在地板上。会议室后面房间的天花板呈拥有凹槽的曲线型，下面还悬挂着一个抽象形状的顶棚。这种流畅而抽象的形式一直延伸到空间内的家具设计中。总经理房间的楼板含有黑色瓦片，与外部起过渡作用的灰色形成鲜明的对比，天花板呈流线型，带有抽象形状的黑色背光板。

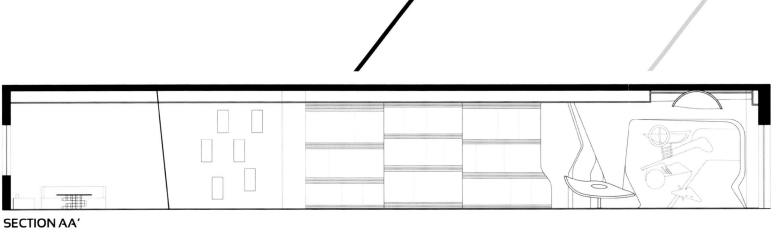
SECTION AA'

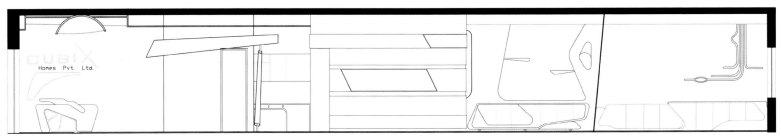
SECTION BB'

SPACES ARCHITECTS@ KA OFFICE

Spaces Architects@ka 办公室

Designer_Kapil Aggarwal / Design Team_Kapil Aggarwal, Pawan Sharma, Chander Kaushik, Karan Arora
Site Supervision_Arvind Pal Singh / Location_New Delhi, India / Area_139.35 m² / Photographer_Bharat Aggarwal

THE OFFICE IN BASEMENT HAS BEEN CONCEPTUALIZED AS AN OPEN OFFICE, THE OFFICE SPACE ON TWO LEVELS, THE LOWER TO BE USED AS WORKSTATION.

COLORS MATCH
色彩搭配

| CMYK 0 0 0 0
| CMYK 18 28 80 0
| CMYK 15 24 48 0

| CMYK 0 0 0 0
| CMYK 40 0 100 0

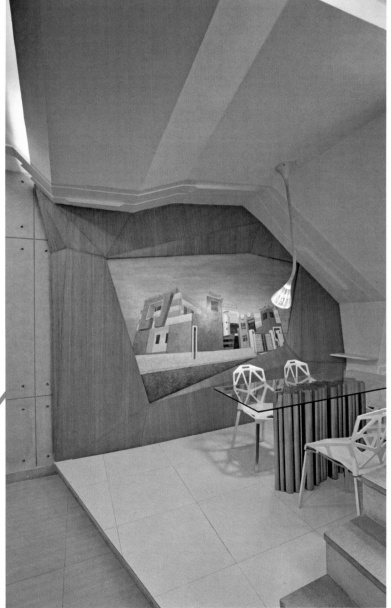

The office design was conceptualized to be place which being leisure is also conductive for people to work in a creative environment, a workplace to enjoy. The zoning of spaces is justified keeping the main cabin with attached conference at the rear to maintain privacy as well as visually connecting it to front office. As the front office space is narrow with regards to rear part, the front space is used as a gallery with walls in cement finish highlighting the project display.

One moves down from front entrance highlighted by fixing glass roof penetrating ample light into the interior space. A raised platform has an informal conference designed by fixing multiple dia. steel pipes with an abstract panel ceiling. The flooring and walls at the front office is kept as cement finish to give emphasis on display panels. An abstract partition acts as waiting, continues on the ceiling extending into multiple abstract boxes displaying different firm's design ideology.

The most interesting part is an experiment with designing of the main cabin outer partition in a fluid form with veneer cladding continuing to the conference room ceiling. The partition is inclined at both the planes and takes an interesting form. The conference and cabin has a glass sliding folding partition which when pulled acts as individual space. A green space with grass flooring and elliptical seating space is used as breakout space in the interior and used for reading books. Two workstations for senior architects are designed behind the seating.

The ceiling plays an important role in the studio creating a visual transition. As the ceiling near the reception made of multiple box panels continued to the ceiling in abstract. Similarly, elliptical ceiling over the reception has a hanging model inspired by Architect's Thesis Project being a focus in space. The circular seating in green area is reflected on the ceiling in an abstract pattern continuing in the rear space.

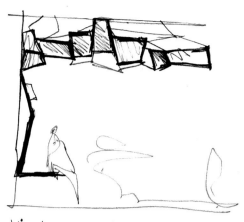

VIEW AT RECEPTION

CEILING AT RECEPTION

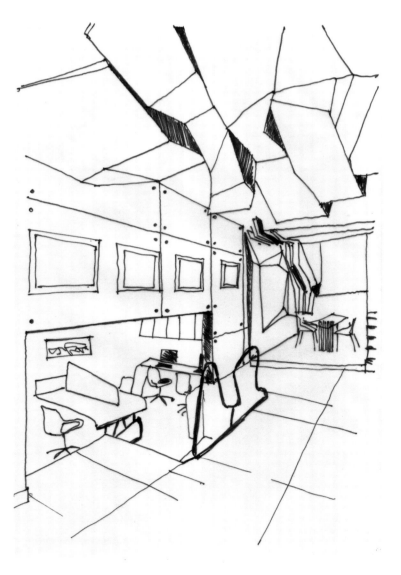

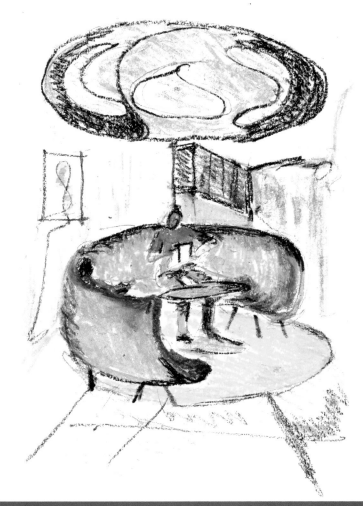

位于地下室的这个办公室被定位为开敞型办公室，该办公室有两层，下面一层是工作站。办公室的设计被定位为工作娱乐兼备的场所，让员工在充满创意的氛围中享受工作。在空间分配上，主要的办公室后面配有会议室以保护隐私，同时，在视觉上与前厅相连接。因为前厅相对于后半部分比较狭窄，所以带水泥墙的前厅被用作画廊来凸显项目、展示产品。

人们从正门入口进入办公室，正门入口是玻璃屋顶，充足的光线透过玻璃照进室内空间。升起的平台上有一个非正式会议室，安装有不同直径的钢管和抽象的天花板。为了突出显示板，前厅的地板和墙壁用水泥粘合剂做成。抽象的隔断被用作等待区，并在天花板上继续延展，形成许多展示各公司设计理念的抽象箱体。

最有趣的部分是对主要办公间的外部隔断的尝试，它是用胶合板包层以流线形式设计的，并一直延伸至会议室天花板处。隔断向两个平面倾斜，样式有趣。会议室和工作室有玻璃推拉活动门，需要时拉上门就是一个独立空间。拥有绿色地板和椭圆形座位的绿色空间被用作室内休息室隔间，人们可以在那里读书。这个座位后面是为高级建筑师设计的两个工作站。

天花板的设计在营造工作室的视觉转换上起着重要的作用。接待处附近的天花板是由箱体面板制作的，以抽象的形式一直延伸到天花板。同样，受建筑师的主体项目启发，接待处上面椭圆形天花板上设计了一个悬挂的模型，成为这里的亮点。绿色区域的圆形座位以抽象的图案反映在天花板上，这种设计方式一直延续到后面的空间设计中。

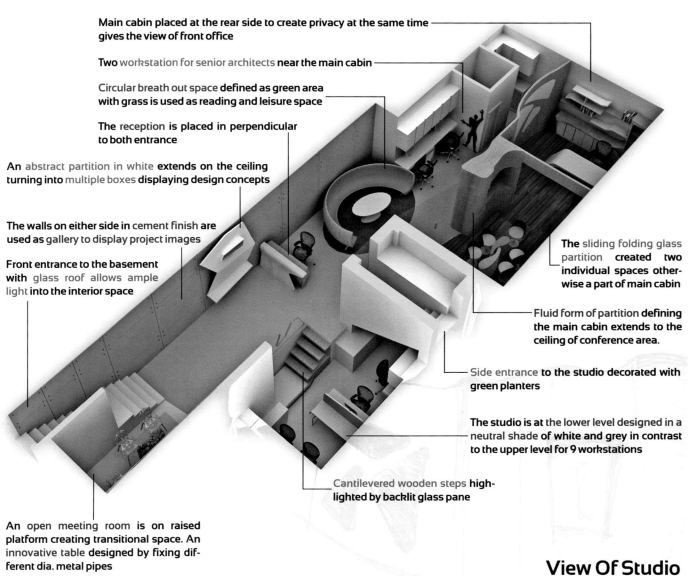

Main cabin placed at the rear side to create privacy at the same time gives the view of front office

Two workstation for senior architects near the main cabin

Circular breath out space defined as green area with grass is used as reading and leisure space

The reception is placed in perpendicular to both entrance

An abstract partition in white extends on the ceiling turning into multiple boxes displaying design concepts

The walls on either side in cement finish are used as gallery to display project images

Front entrance to the basement with glass roof allows ample light into the interior space

The sliding folding glass partition created two individual spaces otherwise a part of main cabin

Fluid form of partition defining the main cabin extends to the ceiling of conference area.

Side entrance to the studio decorated with green planters

The studio is at the lower level designed in a neutral shade of white and grey in contrast to the upper level for 9 workstations

Cantilevered wooden steps highlighted by backlit glass pane

An open meeting room is on raised platform creating transitional space. An innovative table designed by fixing different dia. metal pipes

View Of Studio

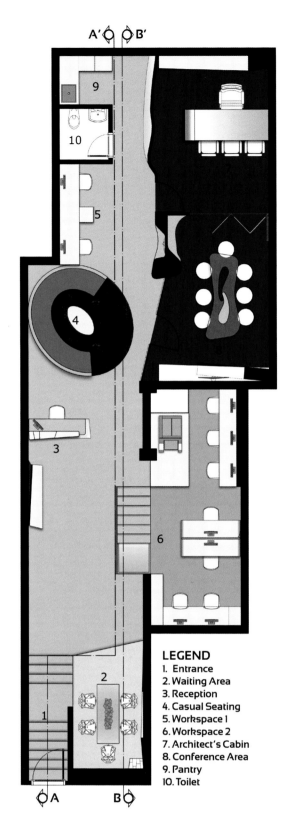

LEGEND
1. Entrance
2. Waiting Area
3. Reception
4. Casual Seating
5. Workspace 1
6. Workspace 2
7. Architect's Cabin
8. Conference Area
9. Pantry
10. Toilet

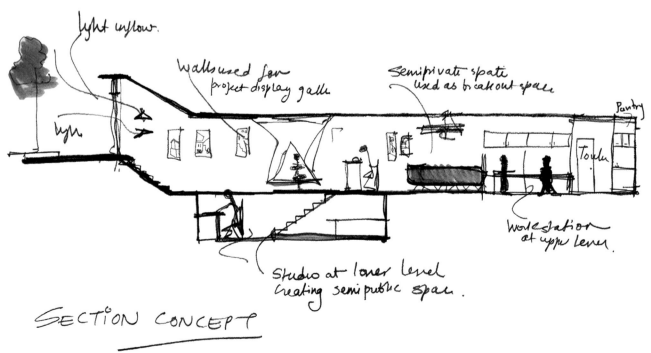

SECTION CONCEPT

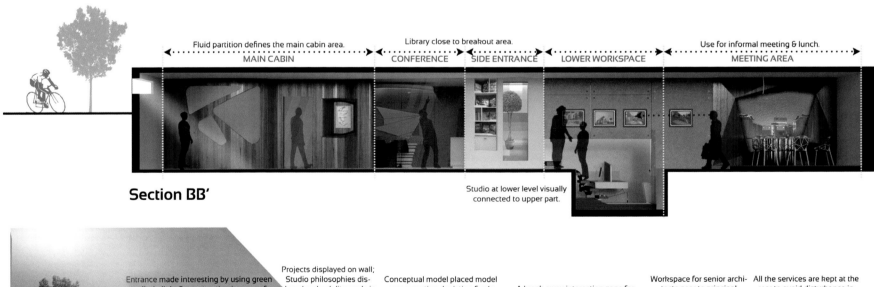

Section BB'

| MAIN CABIN | CONFERENCE | SIDE ENTRANCE | LOWER WORKSPACE | MEETING AREA |

- Fluid partition defines the main cabin area.
- Library close to breakout area.
- Use for informal meeting & lunch.
- Studio at lower level visually connected to upper part.

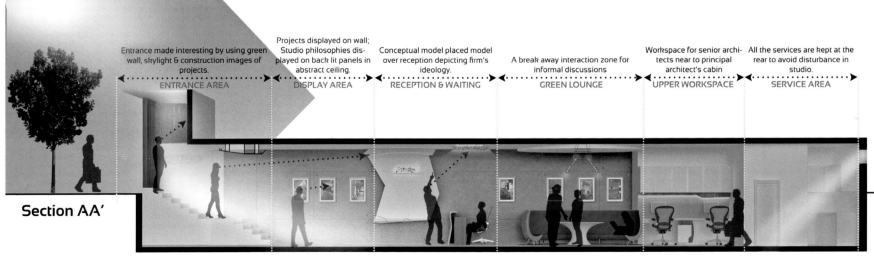

Section AA'

| ENTRANCE AREA | DISPLAY AREA | RECEPTION & WAITING | GREEN LOUNGE | UPPER WORKSPACE | SERVICE AREA |

- Entrance made interesting by using green wall, skylight & construction images of projects.
- Projects displayed on wall; Studio philosophies displayed on back lit panels in abstract ceiling.
- Conceptual model placed model over reception depicting firm's ideology.
- A break away interaction zone for informal discussions
- Workspace for senior architects near to principal architect's cabin
- All the services are kept at the rear to avoid disturbance in studio.

THE BRIDGE

桥

Designer_Threefold Architects / Engineer_Aecom + Tall (Structural)
Client_Bathroom Brands Crosswater / Location_London, UK
Area_1,500 m² / Photographer_Charles Hosea

THE BRIDGE IS A NEW TYPOLOGY OF WORK ENVIRONMENT.

A 64m long, undulating, multi level structure spanning two floors in a double height void within 1500m² of office and recreational spaces.

Conceived as a continuous folded surface, the Bridge is constructed from pre-fabricated cross laminated timber (CLT), and is a structurally dynamic form, spanning over 8m at a time. Bridging between floors, this elegant connecting element encourages interaction between employees and creates pockets of space in which to work and gather.

Both staff and guests enter into a light, airy space where the CLT Bridge structure begins. Here the Bridge creates an dramatic first impression — forming a seating area, a staircase and an impressive 5m high wall, then wrapping upwards to the first floor level.

The Bridge continues to the first and second floors. A connecting element between spaces, it encourages horizontal and vertical movement across the office. This connection is key in addressing the notion of community within the building — bringing together the different departments and companies at strategic points.

Inspired by historic inhabited bridges, the folded CLT structure is sculpted to form spaces above, below and within, for interaction and gathering. These areas for interaction vary in size from 1~2 person booths to a 40 person forum.

Conceived as an extension of the Bridge, here the folded timber surface forms counters, storage and booth areas. The cafe is a light filled space with full height glazing and a large balcony. It is a place for meeting and eating, a place to take clients, and to hold company events.

The rear wall and ceiling above the bridge form the backdrop to the office space. On this blank canvas we created a 48m long installation of gently undulating fins to the wall and ceiling. Rooflights over this space bring in a soft natural light, which is enhanced by a series of long etherial pendant lights. The delicate waves of the fins and lights bring to mind the water that the bridge passes across, creating a dreamlike space within the office environment.

COLORS MATCH
色彩搭配

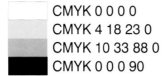

CMYK 0 0 0 0
CMYK 4 18 23 0
CMYK 10 33 88 0
CMYK 0 0 0 90

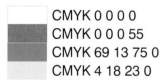

CMYK 0 0 0 0
CMYK 0 0 0 55
CMYK 69 13 75 0
CMYK 4 18 23 0

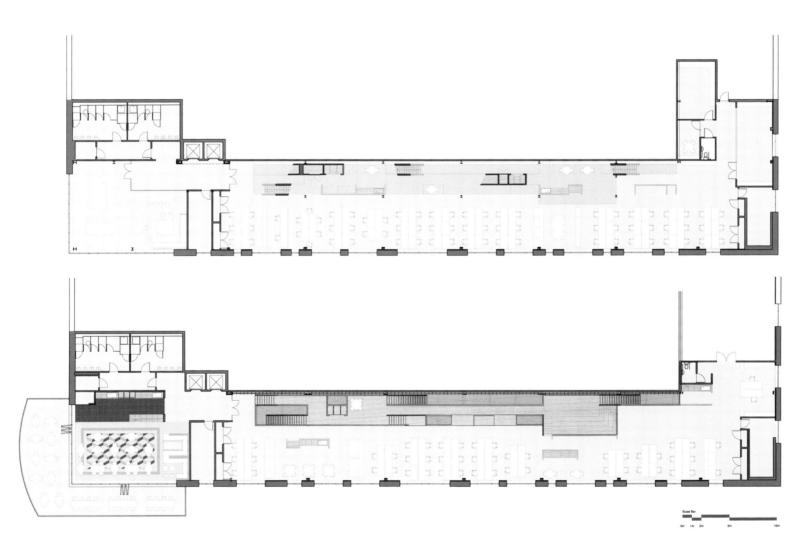
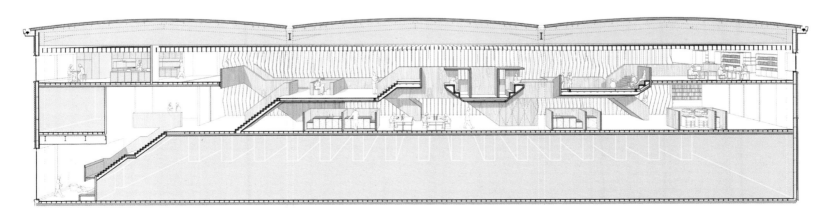

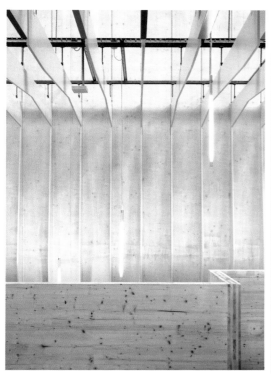

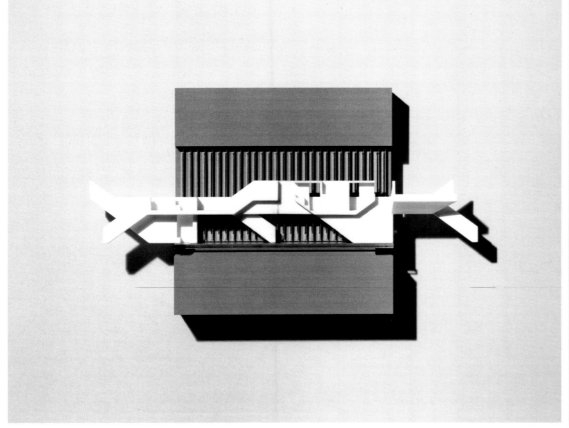

　　桥是一种新型的工作环境，它设在面积为1500平方米的办公和休闲空间里，长64m，呈波浪状起伏，多层次的结构跨越两层。

　　这座桥是由预制的交叉复合木材建成的，表面呈连续折叠状，横跨8m，从而在结构上呈现出一种动态的形式。这座桥连通各楼层，这个优雅的连接元素促进了员工之间的互动，并为员工工作和聚集创造了空间。

　　当员工和客人进入到一个明亮、通风的空间时，交叉复合木材桥结构就会映入眼帘，桥的设计独特，会给人们留下无与伦比的第一印象，它使空间形成一个座位区、一个楼梯和一个令人印象深刻的5m高的墙，并一直向上延伸到二楼。

　　桥从第一层延伸到第二层。空间之间的连接元素促进了办公室内的水平方向和垂直方向上的活动。连接处在诠释"大楼中的小社区"这一概念中起到了至关重要的作用，当讨论战略要点时，不同的部门和公司皆可聚集于此。

　　设计灵感来源于历史上著名的居住桥梁，通过在折叠的交叉复合木材结构上雕琢出不同的造型，从而在桥的上面、下面和中部形成不同的交流和聚会空间。这些区域大小不一，有适合一两个人的狭窄区域，也有能容纳四十多人的会议间。

　　作为桥梁的延伸部分，折叠的木材表面形成吧台、储藏间和展台位。咖啡厅有通高的玻璃窗和硕大的阳台，异常明亮，可以在这里开会和用餐，也可以接待客户，或举行公司活动。

　　办公室空间的背景（后方的墙面和桥上方的天花）长48m，轻轻起伏的鳍片形成墙面和天花板，柔和自然的光线从上面的天窗射进来照在一串悠长而轻飘的吊坠灯上，显得更加唯美。鳍片优美的波动和灯光融合在一起不禁使人们联想到桥下溪水流淌而过的情景，在办公环境中营造一种梦幻的氛围。

INDEX 索引

YUNAKOV ARCHITECTS

Yunakov Architects – The group of designing companies founded and headed by an architect Sergiy Yunakov, the Laureate of the State award, Merit Architect of Ukraine, a corresponding member of the Ukrainian Academy of Architecture, an experienced teacher and architect for more then 30 years. 20-years architectural activity enabled the company to lead the architectural market of Ukraine.

Recently Yunakov Architects presented the new direction – Home Design headed by Ivan Yunakov, which instantly became a successful and popular.

Yunakov Architects creates and keeps archive of scientific and technical information.

To develop project documentation, Yunakov Architects uses present-day achievements and developments in the field of designing and engineering, and in the field of IT-technology as well.

Having strategic project for development and using techniques of project management in work, Yunakov Architects has clear business-process structure, that ensures terms and performance standards required.

Yunakov Architects involves only highly-professional licensed companies for outsourcing, such as MEP-engineering.

Yunakov Architects and constructors of the studio improve their professional skills permanently by studying and assimilating the experience of domestic and foreign design companies.

The field of activity of the company includes designing of residential, office, commercial, hotel complexes, administrative and industrial buildings, and cottage in Kiev and other cities of Ukraine as well.

CACHE ATELIE

Cache atelie (Mila Ivanova and Tsvetomir Pavlov) is young Bulgarian architectural studio that experiments in the sphere of architecture, interior and products design. They are closely involved with developing social art installations and always search for the unexpected and facetious aspects of space.

THREEFOLD ARCHITECTS

Threefold Architects was founded in November 2004 by 3 friends with diverse skills and a shared vision. Over the last 10 years the studio has grown into a team of 9, working on a diverse range of projects. Their designs have won awards and been widely published, and most recently the practice has been shortlisted for the Young Architect of the Year Award 2014.

Threefold Architects believe in collaboration and encourage a strong dialogue between all parties to inform a carefully considered and sensitive solution.

Threefold Architects are interested in people, the way they live, work and play; the way they interact with one another and with the spaces around them, the individual, the pair, the group and the crowd.

Threefold Architects' extensive experience in making private spaces for individuals and families has provided them with an excellent foundation for their recent progression to working with larger groups in the successful creation of multiple living, work and public space.

Threefold Architects' work has won awards, delighted their clients, and been internationally published.

GRAY PUKSAND

Gray Puksand has become known as a pioneering firm in Australia, recognized for excellence in delivering multifaceted projects and visionary designs that emerge from contemporary social, cultural and technological evolutions.

The national firm has evolved into an award-winning, full-service architecture and interior design practice, with more than 100 employees working across offices in Sydney, Melbourne and Brisbane.

A legacy of 150 years of design innovation and problem solving provides their accomplished, multidisciplinary team with a depth of knowledge, skill and agility to produce exemplary designs for clients worldwide.

Every client has different values. Collaboration is integral to their distinctive approach; Gray Puksand's goal is to understand their client's underlying values. Listening, contributing ideas and sharing research knowledge are key aspects of their position as trusted advisors.

SPACES ARCHITECTS@KA

Spaces Architects@ka, an international award winning architectural firm, established with a vision of creating sensible, functional spaces enhanced by the intangible sense of emotion, power and playfulness. This results in architecture that can be extraordinarily responding to the unique needs throughout the design process, is just as much an architect's mission as shaping aesthetically inspired built environment through communication.

Their work is known for its design quality, imagination and originality. Each project offers an opportunity to pursue new solutions to complex building problems. Their motto is to design, detail and enjoy working in responsive to their client's need in very honest, efficient and professional manner. They believe that their works follow a unique style based on their ideology and perspective towards the design requirement.

The word "SPACES" is a very integral part of their being, they move through spaces, they identify with it, their senses response to spaces, they feel it. They work to develop that space for their clients where they emotionally get attached to it; it is not an empty sterile container but a sensual dynamic environment. Their spaces are not formulaic but are derived on the basic requirement of client and site. They feel each house, building or space has got its individual identity, like human no two buildings can be same they as an architect have to give it an identity and soul so they can breathe and communicate. They believe a building is a living object which can communicate; it is just that feeling that has to reciprocate in a very sensitive approach towards their work. Whatever work executed by them one can find that soul or they have at least attempted to and keep on doing that.

The firm in past years has executed more than 100 projects ranging from furniture to corporate buildings with numerous publications.

IPPOLITO FLEITZ GROUP - IDENTITY ARCHITECTS

Ippolito Fleitz Group – Identity Architects is a multidisciplinary, internationally operating design studio based in Stuttgart. Currently, Ippolito Fleitz Group – Identity Architects presents itself as a creative unit of 40 designers, covering a wide field of design, from strategy to architecture, interiors, products, graphics and landscape architecture, each contributing specific skills to the alternating, project-oriented team formations. The firm's projects have won over 200 renowned international and national awards.

JUMP STUDIOS

Jump Studios is a London based architecture and design practice. Established in 2001 by Shaun Fernandes and Simon Jordan, Jump Studios has completed several award-winning projects to date for clients including Nike, Google, Mother London, Levi Strauss, Starwood Hotels, Innocent Smoothies, Red Bull, Adidas, Wieden + Kennedy, The Science Museum and Honda. Jump is currently working on projects for Saatchi & Saatchi, Google, Yahoo! Waze and Rapha among others.

WIRT DESIGN GROUP

Wirt Design Group is a full-service commercial interior design firm located in Los Angeles, CA.

Founded in 1994, the firm has grown into one of Los Angeles' leading commercial interior design practices with a significant portfolio of projects for Clients such as Red Bull, Sempra Energy, Whole Foods Market, eHarmony, Yahoo!, and Northwestern Mutual. To this day, the founding principles of listening to and understanding each client's unique needs still underscore their approach to every project they undertake.

The hallmark of their practice is the cultivation of long-term client relationships and numerous clients have been with WDG for well over a decade. This history and loyalty speaks to their collective ability to listen to their clients and synthesize their needs and goals into an efficient and well-designed workplace.

Wirt Design Group ranks among the city's top commercial design practices and within the top 150 design firms in the United States.

FUNKT ARCHITECTS

Funkt is a group of five recently seven people, and it exists since 2007. They have gathered together to make things, that they care for and they consider that every member of the team is equally important. Professionally they are architects and one marketer, however, they are also involved with product and graphic design, carpentry, drawing, freestyle and everything that could truly motivate them.

They love the simple and unpretentious solutions. They support the traditional crafts and encourage the people who stand up for their revival. They care for the environment and the social impact of architecture, but they try not to take themselves too seriously.

After all architectural projects, private interiors, offices, emblematic public establishments, wooden houses, futuristic visions, furniture and lighting, graphic design, drawings, blogs and causes, they are always interested in something different.

PELDON ROSE

Peldon Rose is one of the leaders in office interior design and fit-out and works closely with their clients to create inspirational and innovative office interiors. Based in Wimbledon, UK and formed in 1986, the Peldon Rose Group Limited is made up on 5 wholly owned operating divisions.

For nearly 30 years, Peldon Rose have completed CAT A and CAT B projects all over the UK, ranging anywhere from 93 m^2 up to 19695 m^2.

Peldon Rose believe in the "client for life" and can provide clients with the right workspace to help increase productivity, attract and retain top talent and bring in the best clients.

ROGERS PARTNERS ARCHITECTS + URBAN DESIGNERS

Rogers Partners Architects + Urban Designers (Rogers Partners) is a New York City based comprehensive, cross-disciplinary studio that focuses on the connective tissue of cities and public spaces. Rogers Partners includes architects, urban designers and landscape architects with deep experience in programmatically rich city-building. Their projects have won more than 60 design and industry awards and have been presented in prestigious exhibitions, including the Museum of Modern Art in New York City. Current projects include the redesign of Constitution Gardens on the National Mall in Washington, DC; Syracuse University's new Energy Campus, Syracuse, NY; the new headquarters for international advertising firm Droga5, New York, NY; SandRidge Energy Commons, Oklahoma City, OK; and Mid-Main, Houston's first transit-oriented mixed-use development, Houston, TX.

3G OFFICE

3g office is an international company of Consultancy specialized in Workplace Innovation, Change Management and Facility Management, with large experience in corporative headquarters of big companies worldwide. They create tailored solutions where customer needs, best practices, and market trends are fit together to deliver a workplace where communication, productivity and employee satisfaction are improved. Their multidisciplinary teams address each project based on three pillars: Spaces, Technology and People, and are experts in Flexible Working and Flexible Office models.

MATT GIBSON ARCHITECTURE + DESIGN

PS ARKITEKTUR

VOA ASSOCIATES INCORPORATED

Matt Gibson Architecture + Design is a Melbourne based design practice that provides architectural, interior, landscape and strategic design services. The work of Matt Gibson Architecture + Design is based upon a solid foundation of design excellence, including budget and programme control, proficient project management and the achievement of best value and architectural quality.

Matt Gibson Architecture + Design is rapidly growing a reputation as one of Australia's best architectural & interior design practices featuring regularly in local and international design publications.

The practice has received a number of awards including recently being shortlisted for World awards for projects the "Kooyong Residence" 2012 and the "Oscar &Wild" fashion retail store 2013 at WAN & WAF Awards respectively. MGA+D has won awards over several categories including residential, corporate and retail from the Design Institute of Australia and the Australian Institute of Architects. The office received numerous other awards recently from the Retail Design Institute 2012, the Australian Timber Design Council 2011, Dulux Colour Award 2012, Melbourne Design Award 2012 and the 2011 Belle House of the Year. In 2009 MGA+D won the World Award for Retail Design presented in Dubai by the International Federation of Interior Designers & in 2005 the inaugural award for Australia's Best Emerging Practice from the Design Institute of Australia.

pS Arkitektur works with projects ranging from commercial interiors to private houses and urban planning. Their vision is to create unique buildings, interiors and environments that make an emotional and visual impression. Their motto is "architecture that makes a difference".

They challenge the obvious solutions and aesthetics in favor of creating something unique, derived from the company's identity and their client's dreams. Their office interiors and refurbishments enhance the clients brand identity and business opportunities.

Architecture and design is a means of market positioning, creating far reaching values for their clients. They define the goals and demands so that the design will be an efficient tool and identity bearer. They bring forward the possibilities, visualize the invisible and suggest changes. They believe that inspiring environments generate positive energy!

In 2011 they received first price for Outstanding Design of the Year at the 9th Modern Decoration International Media Award in Hong Kong, China. Silver in the Swedish award Swedens Prettiest Office, as well as nominations for best office design in Leaf Awards in London and Inside Festival in Barcelona.

Come create with them!

Founded in 1969, VOA Associates Incorporated provides remarkable client service and award winning design for the built environment. They listen to understand client needs and integrate their broad experience to deliver effective and efficient solutions on every project. The result is working, living, learning and healing environments that stand the test of time.

VOA has accumulated extensive experience in healthcare, resort and hospitality design, commercial and mixed-use, multi-family residential, planning, workplace, retail, entertainment and culture, education and government, as well as in not for profit sectors, and has designed a number of acclaimed projects both in the U.S. and abroad. VOA's diversification fosters a vigorous exchange of best practices that continuously results in successful project outcomes.

VOA has been in China since 2006 and in that time they have grown rapidly, having worked in over 25 cities in China, Vietnam, and Cambodia, mainly in hospitality and tourism. They are currently working with all the major International brands in China including Hilton, Intercontinental, Starwood, and Marriott, as well as with Jin Jiang Hotels, one of China's largest hotel operations companies. They are working on resort hotels in Sanya on Hainan Island, Xiamen seaside, Tengchong near the Myanmar border, Beidaihe on the coast in the north and in Anji, China two hours west of Shanghai. They are also working on a number of five star business hotels in Beijing, Changchun and Beidaihe, and have within the last year completed hotels in Shanghai and Beijing.

PEDRA SILVA ARCHITECTS

TKEZ ARCHITECTURE & DESIGN

DEGW

If there is something that summarises who they are and how they work it would be that the best answer always depends on a good question.

At the origin of all their projects is always a unique way of facing a challenge and raising the same question: how do they transform a problem into a unique opportunity?

With more than a decade of experience, they have developed their capacity and flexibility, improved their methods, germinated their experience and know-how.

They grew and they got better and today they're a reference in architecture and interior design, capable of responding to problems by adding value and finding the means to create your dream in any part of the world.

In the mean time they continue to do what they have always done: questioning what makes dreams and transforming an idea or a problem into an object of desire.

TKEZ architecture & design was founded in 2009 by the two partners Tanjo Kloepper and Eugenia Zimmermann.

The Munich based architectural studio develop unique concepts for every single project.

TKEZ generates the quality of their buildings, spaces and finishes by a precise analysis of the specific location, the cultural environment, the program and the requirements and desires of the clients brief.

Currently the studio is working on projects ranging from luxury private residences over retail design to office spaces and furniture design.

The design of concepts for contemporary work environments and office layouts is a field in which the architectural studio has gained special knowledge and profound experience.

Established in 1973 by Duffy, Eley, Giffone and Worthington, DEGW is an international consulting firm specialized in integrated workplace planning. It has been present in Italy with DEGW Italia since 1985. Thanks to an approach based on the research and the observation of organizational behaviours and on how these are influenced by the physical environment, DEGW, for over thirty years, has been able to help companies improving their performance by adjusting the workplace to the corporate strategies and the people's needs.

LABORATORY FOR VISIONARY ARCHITECTURE (LAVA)

ESTUDIO GUTO REQUENA

ROSAN**BOSCH**
ROSAN BOSCH STUDIO

Chris Bosse, Tobias Wallisser and Alexander Rieck founded multinational firm, Laboratory for Visionary Architecture (LAVA), in 2007 as a network of creative minds with a research and design focus and with offices in Sydney, Shanghai(China), Stuttgart and Abu Dhabi.

LAVA explores frontiers that merge future technologies with the patterns of organization found in nature and believes this will result in a smarter, friendlier, more socially and environmentally responsible future.

The potential for naturally evolving systems such as snowflakes, spider webs and soap bubbles for new building typologies and structures has continued to fascinate LAVA – the geometries in nature create both efficiency and beauty. But above all the human is the center of their investigations.

Structure, material and building skin are three areas LAVA believes that architecture can learn so much from nature. Projects incorporate intelligent systems and skins that can react to external influences such as air pressure, temperature, humidity, solar radiation and pollution.

LAVA has designed everything from pop up installations to master plans and urban centers, from homes made out of PET bottles to retrofitting aging 60s icons, from furniture to hotels, houses and airports of the future.

Estudio Guto Requena reflects about memory, digital culture and poetic narratives in all design scales. Guto, 34 years old, was born in Sorocaba, countryside of Sao Paulo State. He graduated as Architect and Urban Planner in 2003 at USP – University of São Paulo. During nine years he was a researcher at NOMADS.USP – Center for Interactive Living Studies of the University of São Paulo. In 2007 he got his Master degree at the same University with the dissertation, "Hybrid Habitation: Interactivity and Experience in the Cyberculture Era".

He won awards in Brazil and internationally. Guto had lectured and exhibited in several cities as New York, Milan, Istanbul, Dubai, Mexico City, Santiago, Paris, Beijing and London. He was a professor at Panamericana – School of Arts and Design and at IED – Istituto Europeo di Design – at both graduation and PHD levels. Guto had lectured on 70 workshops all over the country, and received the "Young Brazilian Awards" as recognition. In 2012 Guto was selected by Google to develop the project for their Brazilian headquarters. And in 2013 Walmart selected him to design their headquarters and in 2014 he won the international award "Building of the Year" from Archdaily with this project under "Interior Architecture" Category.

Since 2012 Guto has a column at newspaper Folha de São Paulo, major Brazilian newspaper, writing about design, architecture and urbanism and also collaborates writing for many magazines. In 2011 Guto created, wrote and hosted the TV show "Nos Trinques", for Brazilian TV Globo channel GNT and developed design web series for the same channel, recorded in Milan, Paris, Amsterdam and London.

Rosan Bosch Studio is a creative art and design studio, creating challenging project that question prevailing norms and conceptions, paving the way for new ways of thinking and acting.

Their projects take their starting point in issues with an overall societal significance as for example how to create a well-functioning workplace, how to make exercise a natural part of the everyday or how to create room for individual, pedagogical development in schools and educational institutions. With a cross-disciplinary grasp in both art, design and architecture, they create completely new solutions and means of development and change – for examples offices with slides and room for play, flexible pop-up workstations and hammocks for individual contemplation or schools with cave-like reading tubes and gigantic icebergs for sitting on as well as teaching in.

As art and design studio, Rosan Bosch Studio is marked by new ways of thinking with an innovative and humane view on design. Through cross-disciplinary projects, the team of designers, architects, academics and artists work to create a societal development – on a smaller as well as larger scale. Common to the projects are that they create environments that make development and change a natural part of the everyday – whether you are learning, working or exercising.

With a focus on creativity, experience and innovation, they translate imagination to physical product and create design and physical environments that make a difference. Their agenda is clear: people should be challenged and developed and design must make a difference!

ZA BOR ARCHITECTS

Denys & von Arend
Interior design - Barcelona
DENYS & VON AREND

za bor architects is a Moscow-based architectural office founded in 2003 by Arseniy Borisenko and Peter Zaytsev. The workshop's objects are created mainly in contemporary aesthetics. What distinguishes them is an abundance of architectural methods used both in the architecture and interior design, as well as a complex dynamical shape which is a hallmark of za bor architects projects. Interiors demonstrate this feature especially brightly, since for all their objects the architects create built-in and free standing furniture themselves. Many conceptual and realized design-projects by za bor architects were awarded at international exhibitions and competitions. At the moment za bor architects is involved in variety of projects in several countries. za bor architects have been involved in more than 60 projects including residential houses, a business center, a cottage settlement, many offices. Among the clients of za bor architects there are IT, media and government companies such as Badoo, Castrol, Forward Media Group, Yandex, Inter RAO UES, Iponweb, Moscow Chamber of Commerce and Industry and others.

Denys & von Arend is a team of recognized professionals and they feel passionate about their work. Their interior design and contracting projects provide overall solutions for hotels, restaurants, offices and showrooms, offering personality, flexibility and made-to-measure budgets. Guaranteed by 25 years of experience, their prestige lies in their clients' trust.

LEMAY

Founded in 1957 as an architectural firm, Lemay is today one of Canada's leading integrated environment design firms, combining architecture, urban design, interior design, landscape architecture, structural engineering and branding into a multidisciplinary, synergetic offer. In the last few years, Lemay has expanded its talent pool by integrating in its rank the firm Eric Pelletier Architectes as well as the three Québec entities from Toronto's IBI Group, which are DAA, Cardinal Hardy Architectes and Martin Marcotte/Architectes. Now well-established in Canada, China, Algeria, Costa Rica and in the Caribbean, Lemay brings together nearly 500 professionals that offer their creativity to support their clients' business strategies.

Lemay — one of Canada's most-awarded firms — delivers projects that achieve the strategic objectives of its clients and exemplify their values, dreams, and aspirations. Drawing on its creative approach and the diversity of its expertise and talent pool, Lemay designs living spaces for tomorrow, constantly striving to build a better future for our communities by creating environments that enhance our quality of life and reduce our ecological footprint. Laureate of numerous national and international awards and contests, Lemay also distinguishes itself by its outstanding leadership, having been recognized as one of Canada's 50 Best Managed Companies by Deloitte.

IND ARCHITECTS

IND ARCHITECTS STUDIO

The studio is featured by a particular attention to details. They believe that it is details that show the quality of architecture.

The team consists of experienced architects who develop the projects starting with a sketch and following it up to complete implementation of intended ideas.

They are united by true passion and commitment.

Since the studio was founded in 2008, they have been dealing with design of apartment and public buildings, town houses and interiors, office premises, hotels, business-centers and restaurants. In these latter days, they are engaged in development of landscape and urban design areas and widely participate in various competitions.

Their strengths: individual attention to every client, effective system of business processes, flexible and quick decision-making, complex approach to architecture and interior, thorough researches and quality analysis at all project phases, quick adaptability to new market requirements.

They provide their clients with comprehensive design documents prepared in accordance to high standards.

Their clients are comprised of those who generate the demand for high-quality architecture: individuals, state officials, businessmen, development corporations.

You can find their works in the most picturesque places of Russia, Spain, Montenegro and Kazakhstan. They are keen to expand this list and take part in up-market international contests.

Looking for substantial cooperation!

M+R INTERIOR ARCHITECTURE

M+R interior architecture is an international operating office founded in 2000 by Hans Maréchal. Their fields of activity often involve complex assignments such as converting and designing offices, airports, libraries, restaurants, hotels, theatres and shops. Among their design skills and core activities for building and interior architecture they are also involved with revitalizing existing buildings and monuments in particular. The architects from M+R determine the form and content of each design assignment on the basis of the programme of requirements. Creativity, functionality, sustainability and ergonomics are translated in a well thought-out manner into a unique final product with an identity of its own. The starting-point is usually a conceptual approach to the design task, for which a total plan is developed, with a sharp eye for details. The power of a strong design is vision, innovation and the quality of realization. They attach a lot of value to the integration of technology and its possibilities. Light architecture is also a specialty within the firm, seeing as light in particular makes an important contribution to the quality of life.

本书的编写离不开各位设计师和摄影师的帮助，正是有了他们专业而负责的工作态度，才有了本书的顺利出版。参与本书编写的人员有：

The people involved in the book are:

Annie Desrochers, Chantal Ladrie, Marie-Elaine Globensky, Véronique Richard, Sandra Neill, Lemay, Isabelle Matte, François Descôteaux, Caroline Lemay, Louis T., Lemay/Technician_Leonor Oshiro, Phary Louis-JeanTechnicians, Serge Tremblay, Geneviève Telmosse, Patella inc., Planifitech Inc., Nicolet Chartrand Knoll Limitée, Claude-Simon Langlois, Eindhoven, Hans Maréchal, Bart Diederen, Marcel Visser, Joris Deur, Adam Smit, Peak Development, Jaap van der Wijst, LBP Sight, Wessels Zeist, Bosman bedrijven, Odico interieurwwerken Deurne, Herman de Winter, Ippolito Fleitz Group – Identity, Peter Ippolito, Gunter Fleitz, Tilla Goldberg, Gideon Schröder, Roger Gasperlin, Tim Lessmann, Trung Ha, Timo Flott, Tanja Ziegler, Lichtwerke Köln, Stefan Hofmann, Zooey Braun, Arseniy Borysenko, Peter Zaytsev, za bor architects, Nadezhda Rozhanskaya, Peter Zaytsev, PTW, LAVA, Simon Parsons, Tony Rossi, Alex Lin, Joanna Varettas, George Freedman, Mark Giles, Chris Bosse, Tobias Wallisser, Alexander Rieck with Jarrod Lamshed, Angelo Ungarelli, Giulia Conti, Brett Boardman, Peter Murphy, etc.

ACKNOWLEDGEMENTS

We would like to thank everyone involved in the production of this book, especially all the artists, designers, architects and photographers, for their kind permission to publish their works. We are also very grateful to many other people whose names do not appear on the credits but who have provided assistance and support, for their contribution of images, ideas and concepts, as well as their creativity to be shared with readers around the world.